ASTRONOMY with BINOCULARS

James Muirden

ARCO PUBLISHING, INC.
NEW YORK

First Arco Edition, First Printing, 1984

Published by Arco Publishing, Inc.
215 Park Avenue South, New York, N.Y. 10003

Library of Congress Cataloging in Publication Data

Muirden, James.
Astronomy with binoculars.

Includes index.
1. Astronomy—Observers' manuals. 2. Field-glasses.
I. Title.
QB64.M86 1983 523 83-7099
ISBN 0-668-05832-3 (pbk.)

Printed in the United States of America

FOR J. H.

Contents

List of Tables

Illustrations

Photographs

Figures

Star Maps

Foreword

When the first edition of this book appeared some fifteen years ago, I could not have anticipated that the use of binoculars in astronomical observation would shortly become so widespread. One basic reason is, simply, that new astronomical telescopes are very expensive. Even a modest 6-inch reflector will cost around $300. This is a considerable outlay to a person who is by no means sure that astronomy is going to become a lasting hobby. On the other hand, there are millions of pairs of binoculars to be found in homes across the country. Even if a pair cannot be found at hand, no great expense is involved; new binoculars of useful quality cost in the region of $30 to $60, and they are something which will always be appreciated by their owner, no matter what course the interest takes.

But something else has happened since 1963: an awareness that interesting and even useful astronomical observations can be made effectively with the relatively very small aperture that conventional binoculars offer. In fact, there are some fields of observation in which they can achieve results denied larger instruments. Astronomy, like all sciences, suffers swings of fashion, and it may be that the current enthusiasm for binocular astronomy is one of those mysterious waves it would be fruitless to rationalize. But, were I asked to pinpoint a single event from which to date this new era, it would be the moment in 1967 when the British amateur astronomer G. E. D. Alcock discovered a nova—an exploding star —using a pair of binoculars.

In the surge of excitement that followed the discovery of the first nova—and in the renewed interest caused by a second—a large number of amateurs took up a pair of binoculars to observe this remarkable star for themselves. It was realized almost at once that elsewhere in the sky were other stars—much less spectacular than a nova, but still varying in brightness—that could be observed using binoculars. These variable stars lay in the no man's land of being too faint for naked-eye observation, and too bright for normal astronomical telescopes; and it was found that many had been neglected for years or decades. The cause was clear: organized binocular work was needed, and the need was met in several ways. In Britain, for example, the Binocular Sky Society, formed in 1968, was soon turning in 20,000 or more observations of these stars annually, and cooperative nova search programs on an international scale were proposed. Further novae were indeed found; Alcock added his fourth in 1976, and John Hosty, using, in fact, one half of an old pair of binoculars, made another discovery the following year. If this enthusiasm is maintained, further finds are inevitable.

This is the background to the present revision, and in updating this book I have constantly borne in mind the increased favor with which binocular work is being viewed. Particular care has been taken to emphasize the fields in which useful work can be done; Chapter 8 has been added for this special purpose. The description of the interesting objects in each constellation has been much enlarged, and coverage has been extended to include the whole sky, with the maps having been redrawn. As far as possible, the constellation notes are based on my own observations; but the descriptions of some far southern groups have had to be taken from other sources. It is hoped that this extension will make the book of use to binocular observers in all parts of the world.

It is a pleasure to record my thanks to Peter York for assisting in the research for this new edition; to the former Director of the

Variable Star Section of the British Astronomical Association, John Isles, for advice on the selection of binocular variable stars; and to the publishers for undertaking the considerable resetting involved.

JAMES MUIRDEN

August 1978

Foreword to the Paperback Edition

In preparing this edition for the press, several of the tables have been updated, and a few illustrations have been changed. A brief guide to the appearance of Halley's Comet in 1985/86 has been added.

Binocular prices have, generally, increased since the last Foreword was written, and even a modest new pair may currently cost between $70 and $100. In contrast, the intensifying competition between telescope manufacturers has kept their prices relatively stable. However, binoculars frequently appear in second-hand lists—and may even be borrowed for nothing at all!

Let me repeat my theme that binoculars are *not* a "poor man's telescope." They show a wide, bright universe, seen not through the constricting lenses of a powerful instrument, but with all the brilliance of two clear eyes, and with the flexibility of a turning head. Even a glance, on a clear night, reveals new wonders: so much more awaits the careful observer with the enthusiasm and the persistence that any worthwhile hobby demands.

January 1983

JAMES MUIRDEN

☆ 1 ☆

Binoculars and Telescopes

During the past few years, with space travel not merely round the corner but actually with us, popular astronomy has been given a record-breaking boost. Almost every week the newspapers announce some astronautical advance, and to cash in on this several manufacturers have mass produced small, cheap telescopes. People buy them, gaze avidly at the Moon for a week or two, and then lose enthusiasm and leave them to collect dust in a corner. They do not realize that the password to amateur astronomy is "patience."

Today, amateurs need even more patience than ever before. Great telescopes are being built to probe regions far beyond the range of smaller instruments, and much work which was truly in the amateur department fifty years ago has been snatched away. But, despite this, well-equipped amateurs can still make useful observations of planets, comets, meteors, and those strange, fluctuating suns known as variable stars.

But not many people have telescopes which will even show a planet's disk, let alone detail on its surface. Does this mean that they cannot make any contribution at all to astronomy? The answer is an emphatic no. Telescopes are useless on aurorae; binoculars are ideal for observing bright comets; and the naked eye can see far more meteors than can be recorded by professional cameras, while it is also the best for observing fluctuations in the brightness of some of the brighter variable stars. All this work is, or can be, of real value.

Of course it is not necessary to do useful work to get satisfaction out of astronomy. Simply watching the changing phases of the Moon, or the lethargic drift of the planets among the stars—best of all, perhaps, simply sweeping the Milky Way with low-powered binoculars; all these make us more intimately concerned with the workings of the universe. We all know that the Moon revolves around the Earth, but it is comforting to check up on successive nights just to make sure. We have seen photographs of the Milky Way which show far more stars than the finest telescope on the clearest night—but once again the secondhand can never match the real. The night sky is full of landmarks, and the best way to get on speaking terms with the stars is by using binoculars or a small telescope. Many amateurs, in fact, know the heavens far better than professional observers.

We do not have to search far for monuments to the humble amateur in present-day astronomy. It should not be forgotten that the vast field of radio astronomy began in 1931 when a "ham" noticed the connection between a recurrent hiss picked up by his radio set and the Earth's rotation. But there have been many sensational visual discoveries, too. For example, on the evening of June 8, 1918, as the sky darkened after sunset, alert amateurs in Europe were amazed to see a bright "new" star shining in the constellation of Aquila; professional observatories were immediately contacted, and productive observations of this exploding star or *nova* were begun. Several novae have been picked up by amateurs during the present century. Another unexpected occurrence was the appearance of a completely new shower of meteors on December 5, 1956, which was well documented by amateur observers in South Africa. Neither need the amateur depend on chance alone: L. C. Peltier, of Ohio, U.S.A., G. E. D. Alcock, of Peterborough, England; J. C. Bennett, of Pretoria, South Africa; W. A. Bradfield, of Dernancourt, Australia; and several Japanese observers, have searched the skies successfully for new comets. Scarcely a year goes by without an amateur somewhere in the world making observations of real importance to professional astronomers.

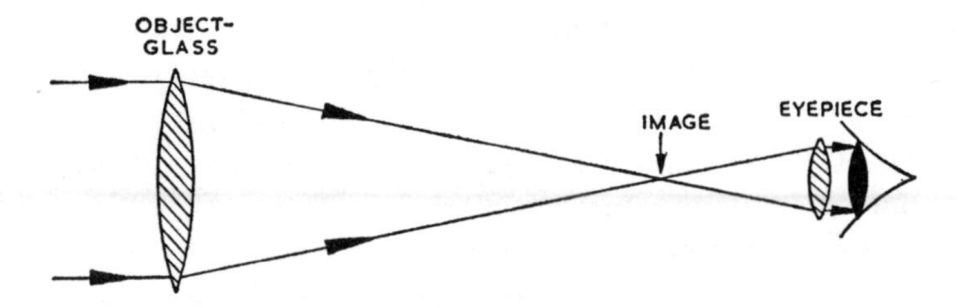

Fig. 1. *A refracting telescope*

The magnification of such a telescope can be found by dividing the distance between the object glass and the image (the "focal length" of the object glass) by the distance from the image to the eyepiece (the focal length of the eyepiece).

Both object glass and eyepiece are shown in their simplest form. In any telescope worthy of the name they consist of two or more lenses close together, the combination producing a better-quality image than a single lens.

The starting point is a pair of binoculars or a small telescope, and they both work on the same basic principle (Fig. 1). Light from the object passes through the large convex lens, known as the "object glass," which bends or refracts the beam, making it converge to form an image. This image is then magnified by a much smaller lens, known as the "eyepiece." It is obvious that the larger the object glass, the more light it will collect; the image will therefore be brighter, or, amounting to the same thing, it will reveal fainter objects. In astronomy, where we are dealing with such dim things as stars, this conclusion is a very important one.

A disadvantage of the telescope shown in Fig. 1 is that it gives an inverted view. Terrestrial telescopes always have additional lenses inside the tube to erect the image, but for astronomical work these are left out; the erector may degrade the image, and it absorbs a slight amount of light—negligible for normal practice, but possibly of great importance when trying to glimpse an excessively faint star. Inversion does not matter at all; drawings of the stars and planets always show south at the top and east at the right, and one sure way of baffling a regular lunar observer is to show him a picture of the Moon the right way up!

For low-power work, however, it is certainly an advantage to have the orientation correct, since telescopic and naked-eye views can easily be compared. In any case binoculars have a system of prisms which effectively fold up a long telescope into a much more compact system, at the same time erecting the image.

Most people think that for any instrument to be at all usable for astronomical work, it must have tremendous magnification. This is a natural enough impression, and it is true that for certain studies (the remoter planets, for instance, or measurements of close double stars) a high power must be used. But no matter what the telescope, a high magnification has two attendant disadvantages. First, the image is made dimmer, because the same amount of light is spread over a larger apparent area, and second, the field of view of the telescope is decreased. It is precisely because they give a bright image and a wide field of view that binoculars are so useful. The brightness of a star, as seen through a telescope, is a function of the aperture rather than the magnification, although magnification plays a secondary part. Therefore, in general, large-aperture glasses will reveal more stars, and may be considered more useful for astronomical purposes. Limits are imposed by cost, and by the general inconvenience of trying to observe with a large and heavy instrument.

Binoculars are always labelled X × Y, X being the linear magnification (e.g., the number of times longer an object looks when compared with the naked-eye view) and Y the diameter of each object glass, in millimeters. The usual conditions are 8 × 30, with a field of view of about 7°, so that fourteen Full Moons would stretch across a diameter. However, they can be obtained to 10 × 50 specification; the field of view is only slightly less (about 5°), while the 50mm object glasses collect nearly three times as

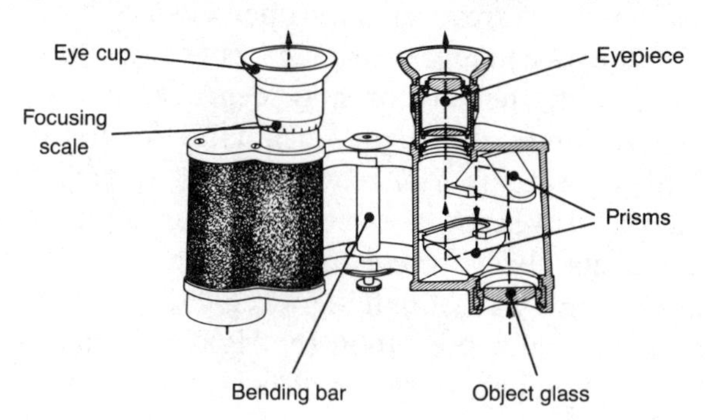

much light. These so-called night glasses are obviously far preferable for astronomical work.

Generally speaking it is not worth buying the very powerful 20 × 60 type; in any case, for the same price one could buy the ingredients for a good reflecting telescope. They are useful for looking at the Moon or for close double stars, but their high magnification means a prohibitively small field of view. It is also a fact that many high-power binoculars are cheaply made, relying on their impressive magnification as a selling point, and will not give satisfactory service.

As far as new binoculars are concerned, there is no such thing as a good, cheap model. A binocular is a complicated instrument, containing a large number of optical components housed in a body that must be light enough to be portable, yet strong enough to house these components rigidly, so that they do not come out of alignment. If the two optical trains of a binocular are not aligned or "collimated" correctly, the views transmitted to each eye will not merge and a double image will result. The eyes have sufficient powers of accommodation to adjust, unconsciously, to a slight defect of this nature, although prolonged observation using such an instrument is liable to give the observer a headache! Serious misalignment cannot, however, be corrected in this way, and makes the binoculars useless until they have been serviced by a specialist.

The drawback of cheap binoculars really centers on their inability to maintain alignment. Other typical defects include the use of poorly-made prisms, which make focused stars look like small crosses through the defect known as *astigmatism,* and bad eyepieces, which make objects appear colored and blurred as they approach the margins of the field. No eyepieces are perfect in this respect, but some are very much worse than others, as will be apparent when binoculars of different quality are compared.

Typical prices for 8 × 30 and 10 × 50 binoculars of reasonable quality are about $70 and $100, and the latter specification is probably the most suitable for general astronomical work. The lower-powered 6 × 30 and 7 × 50 types will give proportionately larger fields of view, but, on the whole, the choice will favor the more powerful models.

Ex-government stores frequently advertise soiled or unissued binoculars at relatively low prices. These are nearly always of excellent quality, and usually have eyepieces with separate focusing adjustments, rather than the center-focus type usually offered commercially. The latter, in which both eyepieces can be focused simultaneously for different distances by turning a central knob, are more convenient for terrestrial viewing; but astronomically the focus is always for infinity, and the individual eyepiece construction is much more rugged—which explains its selection for military use.

Another useful surplus item is the "elbow telescope," which may have an aperture of between 30 and 50 mm, and a magnification of × 7 to × 10. The image is erect, and the eyepiece is at right angles to the main tube, which makes it a very convenient instrument with which to observe objects at a high altitude. Mounted on a light camera tripod with a ball and socket or, preferably, a pan and tilt head, the elbow telescope has the advantage over binoculars that it can be left pointing in the required direction while the observer writes up notes or consults maps.

The question of mountings for binoculars has been the cause of much dispute. It is certainly possible to obtain adequate terrestrial views with hand-held glasses of × 8 or × 10 power, but resting the elbows on a convenient support undoubtedly improves the view. Astronomically, some sort of prop is essential for critical work, for otherwise the stars dance with the observer's heartbeat. A tripod mounting is the fundamental answer, but cannot be used for observing near the zenith; and, in any case, few commercially available tripods are sufficiently tall to allow the observer to stand comfortably at the eyepieces. In my view, by far the best solution is to invest in a reclining chair with arms; if one with an adjustable back can be obtained, so much the better. The observer can then lie comfortably facing the field to be observed, and sweep, if he wishes, through a large arc without having to move. I have spent literally hundreds of hours sweeping the sky reclining in an old fireside chair, using binoculars of up to × 15 power, enjoying both a steady image and easy maneuverability.

Such an observing chair can make life much easier for the observer. Charts for the region being observed can be clipped to an illuminated board in front of him, and can be consulted without having to move the binoculars from the required direction. The importance of observer comfort cannot be overstressed, and nothing is more conducive to dilatory or second-class work than having to go through a good deal of physical effort in order to make an observation. Then again, an orderly way of organizing one's observing procedure works towards the same end. Ideally, it should be possible for the amateur, noting that the sky is clear, to be at work within five minutes, and this can be done if the observing chair, charts and notebooks are all kept handily together. A quick turnout, should the sky suddenly turn favorable, is an essential preliminary to any program.

A useful observing accessory is the old-fashioned opera glass. A pair may be picked up in a secondhand shop for a small amount, and their low magnification (× 2 or × 3) gives a very bright view of the sky. Very large and faint objects, such as the tail of a bright comet, may be expanded beyond visibility with a × 8 instrument, while preserving their form with a very low power. In such cases, an opera glass may be a most valuable aid; its small field of view is its main drawback.

Every astronomer, regardless of the aperture of his telescope, needs a good star atlas. The charts in this book are intended to serve an introductory purpose only. *Norton's Star Atlas and Reference Handbook,* published by Gall & Inglis (Edinburgh), shows stars over the whole sky down to the naked-eye limit, and contains useful tabular information and lists of interesting objects. A more detailed atlas is *Sky Atlas 2000.0*, published by Sky Publishing Corporation, Cambridge, Mass., which shows most of the stars visible with a pair of 8 × 30 binoculars; this will be needed once the observer is seeking the fainter stellar and nebular objects.

The aspect of the star patterns remains the same from century to century, but in front of their background move the Sun, Moon, and planets; while eclipses, comets, and meteor showers perform their periodic phenomena. Details of each year's events are published in many almanacs, including the *Observer's Handbook* of the Royal Astronomical Society of Canada. The world's leading amateur magazine is the monthly *Sky & Telescope,* published in the U.S.A., and the observer will find much of use and interest there. (Details of astronomical periodicals and societies are given in the Appendixes.)

Just occasionally some interesting object, usually a comet, passes through the dawn or sunset region of the sky, and it is difficult to decide whether or not it will be visible against a reasonably dark sky. The answer here is a planisphere, a very simple device which can be set to show the sky for every hour of every day of the year. This is much better than trying to muddle around with one of the monthly newspaper star maps, which do more than anything else to convince people that astronomy is an unintelligible science.

Finally, a word or two about the practical side of observing. Depending on circumstances, it takes the eyes at least ten minutes to get fully dark adapted after going into the garden out of a lighted room; in the brightness the retinal sensitivity is doused, and it is slow to rise to its most sensitive state. For this reason, use the dimmest possible illumination for the map or notebook; a red light affects the dark adaptation of the eyes least of all.

It is interesting, and may be instructive, to compare the two eyes for sensitivity. With binoculars this is not very important, but with an ordinary telescope the situation is different. I have found that my left eye is appreciably more sensitive than my right, and often glimpses otherwise invisible stars. The eyes also see colors slightly differently, which seems to be a common phenomenon.

Always keep a notebook—right from the beginning, even when the observations seem to be of little value. Early drawings and notes will soon be of absorbing interest, and it is an old adage that an observer can always be judged from his records.

☆ 2 ☆
The Sun

The coming of daylight gives no excuse for ending observation. In professional astronomy the Sun occupies a colossal department, for obvious reasons: it is our nearest star, and the only one we can study in detail. While binoculars cannot probe its secrets, they can at least provide an interesting commentary on the lives of its spots.

By our standards the Sun, the center of the solar system, is huge. Yet its 864,000 miles of diameter is not too impressive on the stellar scale; if we turn to Betelgeuse, the red brilliant in Orion, its diameter turns out to be larger than the orbit of Mars. But we can hardly speak of its true diameter, for its surface cannot be properly defined; the gases simply become less and less dense, and like the upper reaches of the Earth's atmosphere eventually fade away into nothing.

The Sun, being smaller and therefore denser (for nearly all stars, regardless of their size, have approximately the same amount of matter in them), has a much more sharply defined surface. We can see this surface, the "photosphere," with the naked eye, and it marks the boundary of the solid, glowing solar matter. In actual fact, however, the Sun's average density is nothing like as great as the Earth's; it is composed mainly of hydrogen, and even though the pressure below the surface is colossal, it is not much more massive than the same volume of water.

Sunspots

The Sun spins like an enormous planet, and we can time its rotation by watching the movement of sunspots—it works out to about 25½ days, though due to the Earth's progress round its orbit it appears to take 27¼ days for a sunspot to return to the same place.

Strangely enough, we still do not really know how sunspots are formed. They are vast, relatively cool areas (the temperature of the rest of the surface is about 6,000° C.); they are strongly magnetic; and they have a powerful effect on terrestrial aurorae. These are caused by electrically unbalanced particles emitted by the Sun affecting the rarefied gas molecules in the upper atmosphere, and whenever a striking aurora is visible there are bound to be sunspots near the Sun's meridian.

Sunspots are certainly within the range of binoculars, though because of the fierce brightness and heat careful precautions must be taken in solar observation. There are two methods: projection of the image onto a screen, and direct observation through a dense filter.

Solar projection

Projection consists simply of holding a white screen a foot or so behind the eyepiece, and focusing the Sun's image on to it; to do this the eyepiece must be racked out considerably beyond the normal position. One of the object glasses must, of course, be covered, and the viewing screen must be shielded from the direct solar rays to prevent hopeless fogging of the image.

While projection is an excellent method when used in conjunction with telescopes of moderate size, when direct viewing is always dangerous, the average binoculars are too low-powered to give a satisfactory image. Certainly it will not reveal nearly so many small features, and it is therefore not to be recommended.

Filter observation

Intense light need not, of itself, be harmful to the eye; it is the radiation normally accompanying visible light which does the damage. For example, the Sun emits radiations of all types, from killing X rays to hot infrared rays (and radio waves as well). Of this radiation, the dangerous type which can pass through the Earth's atmosphere and endanger sight include the barely visible ultraviolet rays, of short wavelength, and the long-wave heat rays. Ultraviolet radiation can destroy living cells, just as heat rays can, but it does not feel "hot." To be safe, then, a solar filter must obstruct these rays as well as tone down the visible light, and it must be placed in front of the object glass, not near the eyepiece where the accumulated heat could burn or shatter it.

Safe proprietary filters of coated plastic (the best known is Solar-Skreen) can be obtained, but a convenient alternative for binoculars is densely fogged and developed X ray film. The crescentic outline of a frosted 100-watt bulb should be barely visible through it. A circle, glued to a close-fitting cardboard cap, can be fixed in front of each binocular objective, and a dim, yellowish solar image, together with any spots that may be present, can be viewed with safety.

Observing sunspots

Binoculars will show a surprising amount of detail in the spots, especially if any of the groups happens to be complex. There are three broad divisions: unipolar, bipolar, and multipolar, depending on whether there are one, two or more main nuclei. The nucleus of a sunspot is its dark center, or "umbra"; surrounding this is a less intense annulus, the "penumbra."

Sunspots come and go, and their lifetimes can be measured in anything from just a few hours, for a tiny transient spot or "pore," to several months. It is interesting to watch a group disappear at the western limb, and to wait for a fortnight to see if it survives the rotation and reappears at the eastern edge. Sometimes considerable groups grow very rapidly, in a matter of a day or two, and these sudden developments are worth watching out for.

Spot sizes are equally diverse. Some are so large as to be visible with the naked eye, when the Sun itself is suitably dimmed, either by an artificial filter or thick haze. Naked-eye spots must be at least 15,000 miles across (this includes the outlying penumbra), but some far larger ones have been observed. The greatest spot group ever known occurred on April 7, 1947; it was bipolar, with an overall length approaching 200,000 miles and an area of 6,000 million square miles, or twenty times that of the Earth!

Sunspot frequency fluctuates in an average period of slightly over eleven years. As yet we have no definite explanation of the sunspot cycle, and neither do we know why it is not completely regular; periods range from nine to thirteen years. The last maximum occurred in 1980–1981, while that of 1957–1958 was the most active ever recorded.

The drop after maximum is much slower than the buildup after minimum; weakest activity was expected in 1975, when binoculars did not show any considerable spot for weeks on end—on the other hand unusual things sometimes happen, a quite considerable group coming into view for no apparent reason, and a close watch should be kept on the eastern limb for any unexpected visitors.

Binoculars will also demonstrate a strange phenomenon known as Spörer's Law. At the beginning of a new cycle spots appear mainly in relatively high latitudes (about 30°), descending towards the equator as the cycle progresses, until after maximum they are mostly grouped in the two 10° latitudes. Around minimum, in fact, we usually have two spotted zones: low-latitude groups from the dying cycle, and high-latitude spots heralding the arrival of the next one.

Strangely enough, spots hardly ever appear very near the equator, and never near the poles; the most arctic spot ever recorded was in latitude 65°.

Accuracy in making daily drawings will come with practice, but there should certainly be no difficulty in detecting the solar rotation from day to day. Keep a special eye on the tiny spots near the limit of visibility, for these are breaking-out grounds for future groups—and remember that a great deal can happen to a region while it is on the averted hemisphere!

Careful observation over a number of months will show that the

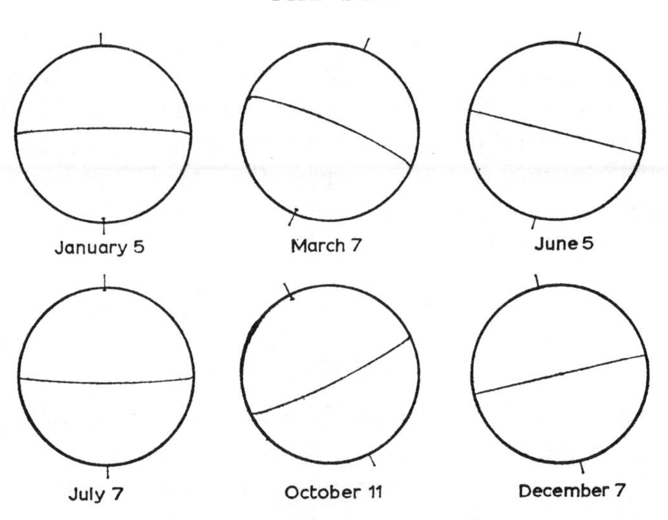

Fig. 2. *Different views of the Sun's axis*

On January 5 and July 7 the solar axis appears vertical (i.e. north and south). On March 7 and October 11 the north and south poles respectively are at their maximum presentation. On June 5 and December 7 the Earth passes through the plane of the solar equator.

The drawing gives the erect view, with north at the top. The Sun spins from left to right.

sunspots' apparent motion changes (Fig. 2). This is because the Earth's orbit lies at an angle with the solar equator. Starting off in June, when we lie in the equatorial plane, they seem to move in straight lines from left to right. After that we drift slightly north, and the south pole disappears; the paths therefore appear curved. The maximum effect is in September, after which they straighten; in December the south pole reappears and the north moves off the disk, when the paths curve in the opposite direction.

For the same reason the solar axis appears to swing first one way and then the other. In July and January it is vertical. The greatest western tilt of the north pole ($26\frac{1}{2}°$) is on April 8; the corresponding eastern tilt, on October 11. Due allowance must be made for this when drawing in a north-south meridian line on the disk, though with binoculars the slant can be estimated only roughly. In any case the tilt from the vertical is true only at noon, when the Sun is due south; when rising its axis slants over to the left, and when setting it leans to the right.

Sunspots appear black, but this is simply due to contrast with the brilliant photosphere; they are really hot and glowing, though much cooler than the rest of the surface, and if they could be seen by themselves they would shine brightly. This is best seen during a solar eclipse, when they appear distinctly brown compared with the black lunar disk.

Solar eclipses

A solar eclipse occurs when the Moon passes in front of the Sun. This can obviously only occur at New, when the two are more or less in line anyway; if the Moon happens to pass exactly through the plane of the Earth's orbit, it will either partly or totally block out the brilliant solar disk.

Fortunately for us, the Moon appears about the same size as the Sun—the Sun is 400 times bigger, but it is also 400 times as far away—which means that at a central eclipse it is usually completely covered. Not always, however; the lunar orbit is not perfectly circular, and just occasionally an eclipse occurs when its distance has shrunk it too small. In this case the Sun's boundary appears as a complete ring, and the eclipse is called "annular."

Because the Moon only just covers the Sun, the region of totality on the Earth's surface for any particular eclipse is very restricted; the shadow is about a hundred miles wide, stretching eastwards as the Moon's orbital motion sweeps it over the surface. This means that the chance of any particular place witnessing a total eclipse is about once in 360 years, and the would-be observer must be prepared to travel a considerable distance in order to observe one.

The partial zone is, of course, far wider, extending for thousands of miles; and any given site is likely to fall within the partial shadow of an eclipse every two or three years. The table on page 15 gives details of the visibility of forthcoming solar eclipses; eclipses occurring in summer are, for obvious reasons, likely to enjoy better observing conditions.

Binoculars can provide interesting views of partial eclipses. First of all there is the noting of "first contact," which is the

Table 1 *Forthcoming Solar Eclipses*

Date	Eclipse
1983	
Dec 4	Annular; track from the Atlantic Ocean, across central Africa, into Somalia
1984	
May 30	Annular; track from the Pacific Ocean, across Mexico and the southwestern USA, across the Atlantic Ocean, into Libya
Nov 22/3	Total (maximum duration lm 59s); track from the East Indies and across the southern Pacific Ocean
1985	
May 19	Partial (maximum phase 0.84, in the Arctic)
Nov 12	Total (maximum duration lm 55s); track from the southern Pacific Ocean into Antarctica
1986	
April 9	Partial (maximum phase 0.82, in Antarctica)
Oct 3	Annular; a short track in the northern Atlantic Ocean. A total phase lasting for perhaps one second may be seen from the midpoint of the track!
1987	
Mar 29	Total (maximum duration 0m 56s—annular at the ends of the track); track from Argentina, across the Atlantic Ocean and central Africa, into the Indian Ocean
1988	
Mar 18	Total (maximum duration 3m 46s); track from the Indian Ocean, across the East Indies, into the Pacific Ocean
Sept 11	Annular; track from the Indian Ocean, south of Australia, into Antarctica
1989	
Mar 7	Partial (maximum phase 0.83, in the Arctic)
Aug 31	Partial (maximum phase 0.63, in Antarctica)
1990	
Jan 26	Annular; track confined to Antarctica
July 22	Total (maximum duration 2m 33s); track from Finland, across the USSR, into the Pacific Ocean

instant at which the Moon's eastern limb begins to cover the Sun; a publication such as the *Observer's Handbook* gives the time, and also the "position angle." Position angle is reckoned in degrees, starting from north (twelve o'clock), round by east (90°, nine o'clock), south (180°, six o'clock), and west (270°, three o'clock) (Fig. 3). Since the Moon first touches the Sun on its western limb, the P.A. of first contact is somewhere around 270°. The ever-widening segment should be visible after a few seconds. Keep a lookout for any sunspots that may be occulted by the lunar disk.

The duration of a partial eclipse, from first to fourth or last contact (the time at which the Moon moves completely off the Sun), depends on its phase. For example, the 5% eclipse of November 23, 1965, as seen from Sydney, lasted for only 49 minutes, while the 96% eclipse observed from Perth on June 20, 1974, had a duration of 1 hour 51 minutes.

The longest possible duration of totality is only about $7\frac{1}{2}$ minutes, and most eclipses are much shorter. One of the longest eclipses of recent times occurred on June 30, 1973, but poor atmospheric conditions hampered most observations.

Anyone with a chance of observing a total solar eclipse will find a pair of binoculars to be the ideal equipment. For observing the corona and the rosy *prominences,* red clouds of glowing hydrogen that rise from the solar surface and appear in profile at the lunar limb, the dense filters are, of course, removed, since the radiative intensity of these phenomena does not much exceed that of a Full Moon. But herein lies the risk that the observer may catch a glimpse of the narrowing solar crescent. The caps must, therefore, be left on until the last moment before totality.

I still have vivid memories of the total eclipse of September 22, 1968, which was observed from a station in Siberia. This eclipse

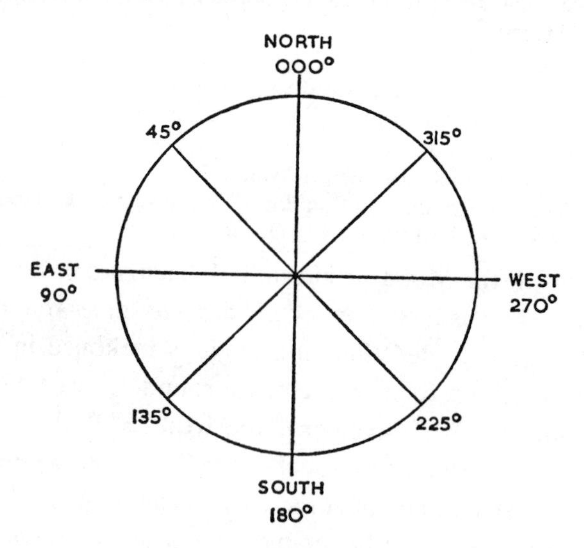

Fig. 3. *Position angle*

was, perhaps, unique in that all the expeditions from different countries had been brought together in the same meadow; and a happy international event it turned out to be. An hour before totality, the Moon made its first nick at the right-hand solar limb. Before long, the Sun was reduced to a thick crescent, and the illumination of the landscape had become noticeably paler; ten minutes before totality the narrowing solar disk was throwing sharp, dark shadows, and the light had a yellow quality, as if a theater spotlight was shining down on the scene.

Quickly, now, the light faded. Only the thinnest silvery crescent remained; and then, as it broke up into points of light, the *Baily's Beads* caused by fragments of sunlight shining momentarily through valleys at the Moon's eastern limb, the red prominences shone out. As the beads died, so the petallike corona blossomed around the black lunar disk. Through binoculars, the blue-white corona, turning almost to turquoise at the lunar limb, and the plum-colored prominences shining against it, formed an unforgettable sight. For 37 seconds we were treated to this glimpse of the Sun's hidden splendors; and no one disputed that they had been worth the many thousands of miles' traveling involved.

Other solar phenomena

When the Sun is very near the horizon our atmosphere plays strange tricks with its shape. The reason here is irregular refraction of light. When very low in the sky a star appears to be slightly higher than it really is; at an altitude of 10° the effect is about $\frac{1}{10}°$, but it increases to $\frac{1}{2}°$ at the horizon. This means that a star or planet can actually be visible above the horizon after it has set, or before it rises. The Sun and Moon are both $\frac{1}{2}°$ across, so that when they appear to rest on the horizon they are really below it.

The Sun's lower limb being more affected than the upper one, the effect of refraction is to squash the disk into a steadily more eccentric ellipse. Sometimes, however, remarkable distortions occur; the limb may become serrated, and on occasions the upper part of the disk appears separated from the lower. This happens when there are markedly divided heat layers in the atmosphere, since the degree of refraction depends on the temperature of the air the light is passing through.

Anyone who is fortunate enough to have a very low, preferably sea, horizon, can look out for a fleeting phenomenon known as the "green limb." It usually occurs just as the last fragment of the solar disk lingers on the horizon, when for the fraction of a second the color changes from yellow to a bluish green. Apart from waiting for a very clear sky, which is essential, the main secret is not to look at the Sun until the last moment; from a point of view of avoiding dazzling the eyes, conditions are rather more favorable at sunrise provided the exact point of appearance is known beforehand. The tint itself is caused by a combination of refraction and absorption in the atmosphere. I have seen this phenomenon, using binoculars, several times. Far more rare, though usually confused in name, is the "green flash," which takes the form of a vivid ray shooting from the sun's upper limb towards the zenith at the moment of sunrise or sunset. This effect is associated with extreme atmospheric turbulence.

☆ 3 ☆

The Moon

Strangely enough, our nearest neighbor is slightly frustrating to the binocular observer. It is near enough to be tantalizing, but not sufficiently close to show interesting detail without a fairly high power. It is here that the relatively powerful 15 × 40 or 20 × 60 binoculars come into their own, for the Moon has plenty of light to spare and the smaller field is, of course, no disadvantage at all.

Our one natural satellite has been in the publicity spotlight for some time now, and the basic facts are well known. It is a much smaller world than the Earth, with a diameter of only 2,160 miles. In addition it is much less dense (it can have only a very small metallic core), so that its mass is less than $\frac{1}{80}$. The surface gravity is therefore considerably less, which means that it has been unable to hold on to any appreciable atmosphere at all.

The Moon revolves around the Earth at an average distance of 238,000 miles. This actually varies from 226,000 miles at its closest point, or "perigee," to 252,000 miles at "apogee," since its orbit, in keeping with all orbits in the solar system, is not perfectly circular. The difference is, however, so slight that for most purposes we can neglect its eccentricity.

The Moon's phases

Fig. 4 explains the lunar phases. At position A the Moon is very close to the Sun in the sky, and the night side is turned towards us; it is therefore invisible. This is New Moon. Gradually it moves towards B, appearing first as a crescent in the evening sky and finally as a perfect half—this is First Quarter, which means that it has achieved one quarter of its total journey round the Earth. Each quarter takes about a week.

As the "terminator" (the line separating day from night) becomes convex instead of concave, we enter the gibbous phase. All this time the Moon is setting later and later after the Sun, since it is getting farther away; by the time it reaches C the entire daylight hemisphere is turned towards us, it appears as a perfect circle, and it is opposite the Sun in the sky. This is Full Moon, when it rises at sunset and sets at sunrise. If the lineup is absolutely perfect there will be an eclipse, but we shall see presently that these are comparatively rare.

After Full the terminator passes to the opposite, or eastern, side. The phase begins to shrink, and the Moon rises late in the night when Last Quarter (D) is reached. After that it is visible as a crescent shortly before dawn, and after 29½ days it is back at New again. This period is the lunar month.

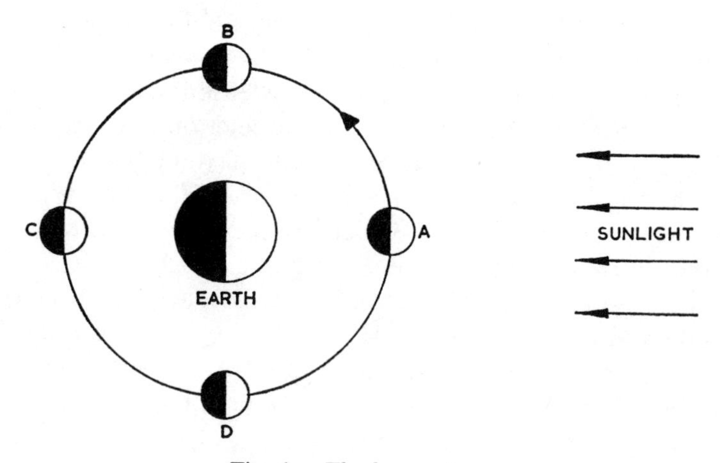

Fig. 4. *The lunar phases*

Oddly enough, the Moon takes only $27\frac{1}{3}$ days to make one revolution around the Earth when measured relative to some stationary object, such as a star. This is because the phases depend on the position of the Sun. The Earth, as it moves along its orbit, carries the Moon with it; the direction of the sunlight is changing all the time, and to make up for this the Moon has to describe slightly more than a true revolution.

The surface features

Plates 1 to 4 show the Moon at various phases. Notice, in particular, how the craters and mountain chains stand out along the terminator, where the Sun is still low in the lunar sky. Every feature casts its own private shadow, which has disappeared by local midday; the region then appears as a featureless, glaring mass of light. Observation of any particular feature is therefore most spectacular when it is seen during local morning or evening.

A glance at the Earth from a point out in space would reveal two striking divisions of the surface: the land and the sea. We can divide the Moon into the same two regions, though perhaps less accurately; the great dark areas, which are relatively smooth, are the seas or "maria" (actually consisting of solidified lava, thinly overlaid with dust), and the glittering rugged uplands are the land proper.

The craters which honeycomb the uplands are the characteristic lunar features. Basically they are huge saucerlike depressions, with rugged walls and frequently a colossal mountain group at the center. Binoculars will show hundreds, especially in the southern hemisphere, and some are so big that smaller editions of themselves may be glimpsed on their floors. The 150-mile Clavius (see the outline map), near the south pole, is a good example.

When the newly discovered telescope came into general use astronomers spent years puzzling over the problem of crater naming. Eventually the solution was found by an Italian priest named Riccioli, who published a map in 1651 on which he had called them after famous scientists and philosophers in general; it was a good idea because it served to keep their names constantly in mind, and also, as new craters came to light, so new names were available with which to decorate them. Nowadays they range from the ancient Greeks to contemporary astronomers, amateur as well as professional.

Due to the importance of having the terminator passing through the region under study, it is impossible to survey all the interesting features on just one night. With this in mind, the following remarks apply to the different phases from New to Full.

The thin crescent

When it first becomes visible in the evening sky one large feature dominates the crescent: the elliptical Mare Crisium, very near the eastern limb. Approximately 300 miles across, it is one of the smaller maria; nevertheless it is probably the most interesting of all. Binoculars will just reveal its mountainous border, and the best time to look at it is actually just after Full, under evening illumination. In a powerful telescope the apparently smooth surface of the mare is seen to be pimpled with tiny craterlets a couple of miles across, winding ridges running their way among them.

The Mare Crisium is a useful guide for revealing the lunar "libration," or apparent slight swinging of the Earth-turned hemisphere from east to west and back again. The Moon's orbit is slightly elliptical, and from the universal laws of motion it follows that the closer a satellite is to its parent planet, the faster it must travel in its orbit. When the Moon is near perigee it therefore speeds up, and near apogee it slows down. Meanwhile, its axial rotation has remained steady. The result is that the two rotations periodically get out of step. At one observation the Mare Crisium may almost touch the limb, while twelve days later it will have swung well clear. This slow shift, once it is known to exist, can be easily detected with the naked eye.

Down south of the Mare Crisium is a prominent line of four craters: in order, Langrenus, Vendelinus, Petavius and Furnerius. They are all eighty to one hundred miles across. Vendelinus, on the shore of the Mare Foecunditatis, has been broken down by the once molten lava—one of the many lunar Pompeiis to be seen all over the surface. The most interesting is Petavius, which has a fine central mountain and a deep valley running to the southeast wall.

Another conspicuous crater is Cleomedes, just north of the Mare Crisium.

The crescent Moon is highest in the spring, and when it is well placed the dark side can nearly always be seen, glowing with a gray light; it is frequently very obvious with the naked eye, in which case binoculars will show some vague detail in the night hemisphere. The phenomenon, known as "Earthshine," is caused by sunlight reflected from the Earth relieving the stark blackness of the lunar night. Our home planet must glow brilliantly in the star-studded sky, especially when there is a dense cloud cover, and Earthshine observations give some sort of clue to atmospheric conditions on our sunlit hemisphere.

Four days old

The terminator now passes through the Mare Tranquillitatis, and a fine chain of craters leads off its southern border: Theophilus, Cyrillus and Catharina. Lines such as these were often cited as evidence for an internal origin of the lunar craters. But, once samples of the surface were examined after the Apollo missions, it became clear to most people that these features are the scars of impacts with interplanetary bodies several miles across. The surrounding surface shows clearly the marks and remains of material blasted out from the subsurface layers when these impacts occurred some 4,000 million years ago.

Certainly there are also volcanic features, such as the Hyginus Valley near the center of the Moon. But the scale of these formations bears no comparison with the size of the impact craters, and the "chains," so called, seem to have no more significance than the random alignments to be expected in any chaotic bombardment.

In the southern hemisphere craters crowd the terminator and must be identified by means of the outline map. Of more interest is the group of three craters near the north pole: Atlas, Hercules and Endymion. In a large telescope both Atlas and Endymion are seen to have dark patches on their floors, and some observers consider them to vary in extent and tone during the lunar day. Others oppose this view, and it is certainly hard to account for any such change.

First Quarter

Probably the most spectacular view of all is obtained when the Moon appears as a perfect half. The first thing to catch the eye is the great string of craters running down the central meridian. There are six main ones: Ptolemaeus, Alphonsus and Arzachel; below them Purbach, Regiomontanus (a chart-confusing name) and Walter.

The 90-mile Ptolemaeus is a magnificent object, and luckily for us it is almost exactly at the center of the disk. In binoculars its floor seems featureless, but it really contains a great deal of fine detail; there is also a conspicuous craterlet Lyot (the eminent late French astronomer). Its mountain ring, however, is nowhere very high, and soon after the Sun has risen sufficiently high to douse the shadows its outline becomes very obscure. A neighboring crater, Hipparchus, slightly smaller but once just as magnificent, has been so battered by the Moon's early geological upheavals that it is hardly recognizable at all except when close to the terminator.

In the northern hemisphere the Mare Serenitatis is well placed. Its southeast border merges with the Mare Tranquillitatis, while its western shore is defined by two magnificent mountain ranges, the Caucasus and Haemus Mountains. Due north of the Caucasus are the Alps. These pay host to the curious Alpine Valley, which cuts so clean through the mountains that it looks as though some gigantic body just wiped the peaks out of its way. The "grazing meteorite" theory can hardly be taken seriously, and it must be due to a surface collapse.

Worth mentioning, even though it is invisible with binoculars, is the craterlet Linné in the Mare Serenitatis. Before the middle of last century it was seen and drawn as a distinct crater; now it appears as nothing more than a tiny pit at the center of an extensive white area, and a comparison between appropriate maps shows a very marked contrast indeed.

Nevertheless, the Moon can play strange tricks. The look of a region changes amazingly from night to night as the sunlight strikes it from a steadily changing angle, and the Linné case could just possibly be put down to a remarkable coincidence of errors on the part of the earlier observers. At all events, it was the most concrete evidence of lunar activity until the Soviet astronomer N. Kozyrev detected gaseous carbon dioxide near the central peak of Alphonsus in 1959; this, at least, is quite definite.

Ten days old

It is a good idea to look at the Moon just after First Quarter, for this is the time when a magnificent mountain range can actually be seen with the naked eye. The mountains concerned are the Apennines, which sweep down across the central meridian. If the phase is just right their sunlit peaks jut out over the terminator, and for a few hours the little projection of light is easily visible.

The Apennines form the eastern border of what is probably the finest mare of all: the Mare Imbrium. It is now well in view; small craters are dotted here and there, but especially noticeable is the group of three in the shelter of the Apennines: Aristillus, Autolycus and Archimedes. It is somewhere in this vicinity that the first lunar missile, *Luna 2*, landed in 1959.

Ranking with the most interesting craters on the entire Moon is the dark-floored Plato, on the northern shore. Regular in outline, its 60-mile floor is almost featureless in small telescopes. In larger instruments, however, a number of tiny craterlets appear, whose visibility seems to be strangely variable; it is just possible that occasional very slight hazes cover this and other regions on the lunar surface.

The finest crater of all, Copernicus, lies in the Mare Imbrium. It is a perfectly formed crater, 55 miles across, with walls rising in places to 17,000 feet above the inner floor; in the center is a superb mountain group, the central peak reaching to 22,000 feet. In a large telescope, when the lighting is just right, Copernicus is a wonderful object. It lies on a plain slightly elevated above the general level of the surrounding mare, and it is the center of the second-largest ray system on the Moon.

Rays, the white streaks which radiate from certain craters like the spokes of a wheel, number among the many lunar mysteries. They are often hundreds of miles long, and the most prolific ray-center, the 50-mile crater Tycho in the southern hemisphere, has given birth to a couple of streaks which extend for a thousand

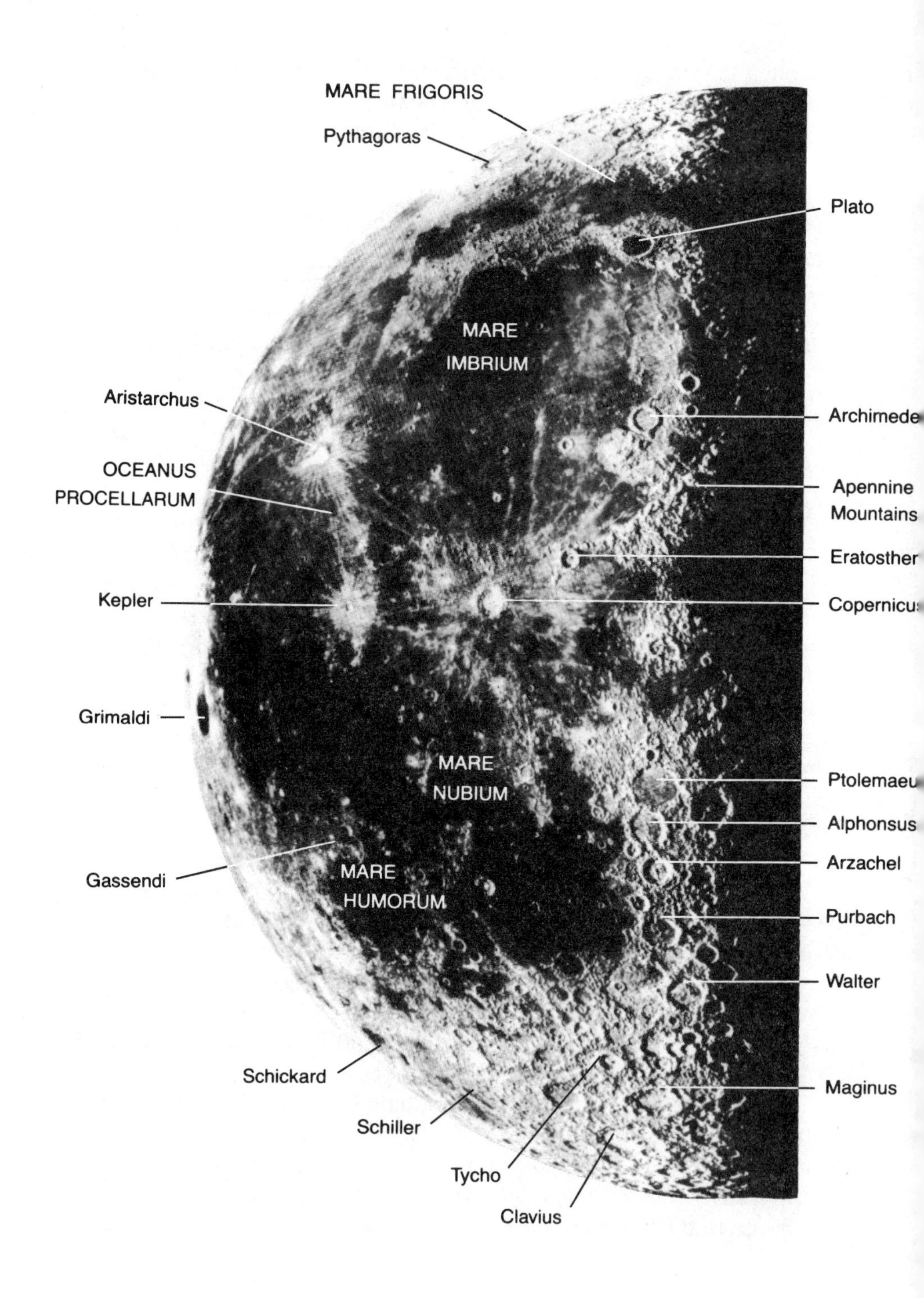
MARE FRIGORIS
Pythagoras
Plato
MARE
IMBRIUM
Aristarchus
Archimede
OCEANUS
PROCELLARUM
Apennine
Mountains
Eratosther
Kepler
Copernicu
Grimaldi
MARE
NUBIUM
Ptolemaeu
Alphonsus
Gassendi
MARE
HUMORUM
Arzachel
Purbach
Walter
Schickard
Maginus
Schiller
Tycho
Clavius

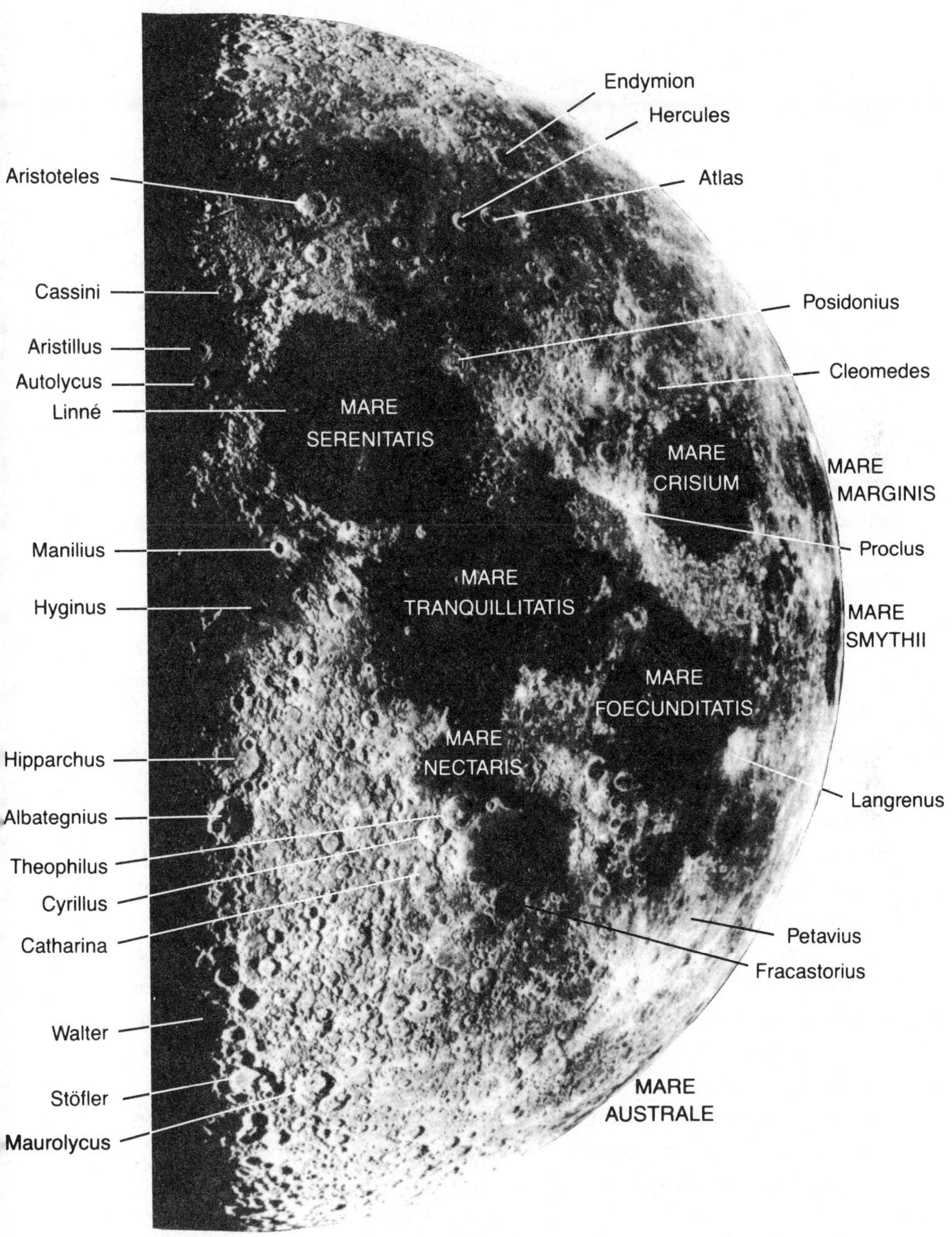

Fig. 5. *Map of the Moon*

miles over the rugged, bleak landscape. Rays are best seen at midday (at Full they dominate the disk), and they remained a puzzle for many years. We now think that they must be due to fracturing of the surface when these very late crater-forming impacts occurred. Through these fractures crept the glasslike material from beneath the crust that now reflects the sunlight so well.

Another lunar landmark is the beautiful Sinus Iridum, or Bay of Rainbows, on the northern shore of the Mare Imbrium. It is a semicircular cliff formation over a hundred miles across, looking very like the remains of a once noble crater that had its southern wall destroyed, and its floor flooded, by the relentless boiling lava of the young mare. When the phase is right the cliff juts over the terminator in a scimitarlike curve; a fine sight even in binoculars.

Somewhat south of Tycho is the most impressive crater of all: Clavius. It is 150 miles across and beaten into second place by the nearby Bailly, which has a diameter of 180 miles—but Bailly is a rather obscure object, close to the limb. Even binoculars will show a string of quite considerable craters spread across the floor of Clavius.

Full Moon

Near Full the brightest feature of all comes into view; this is the 23-mile crater Aristarchus, in the Oceanus Procellarum. All the uplands on the Moon are overlaid with a gray deposit, which presumably is meteoric ash, but the coating on Aristarchus is certainly unusual. It is one of the features which often show up distinctly under Earthshine conditions.

Remarkable contrast is afforded by the nearby Grimaldi, a colossal crater 120 miles across, but so near the limb that we get only a very oblique view of its floor. This floor is of a dark steel-gray hue, even darker than that of Plato, and with the exception of one or two isolated spots it is the darkest surface on the Moon. Like the Mare Crisium, Grimaldi is a good guide for libration conditions.

Full Moon is, in general, the worst possible phase to observe; the sunlight is striking the surface from behind our backs, so to speak, and shadows are at a minimum. Just occasionally, however,

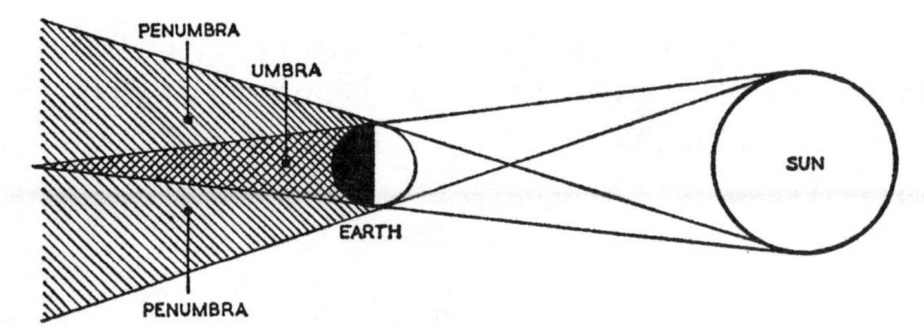

Fig. 6. *Lunar eclipses*

In the dark central cone, the umbra, all direct sunlight is cut off. This merges into the penumbra, where the Sun is progressively less and less obscured. In this region an observer on the Moon would see a partial solar eclipse.

Full occurs almost exactly in the straight line from the Earth to the Sun, and when this happens the sunlight is cut off for a time and we see a lunar eclipse.

Lunar eclipses

The Earth casts a tapering shadow in space, a shadow which at the Moon's mean distance is just over 5,700 miles wide. Moving at an average speed of nearly 2,300 miles an hour, our satellite can therefore remain totally eclipsed for up to 1 hour 40 minutes. Add to this the time taken for the phase to become total, as well as that spent in the lighter "penumbra" of the Earth's shadow (as opposed to its darker core, the "umbra"; Fig. 6), and the total duration of a lunar eclipse can be over six hours.

The Moon, however, never disappears completely when it passes into the umbra; the shadow is lit up by sunlight refracted inwards by our atmosphere. Red light is refracted most of all, which explains why the eclipsed Moon is usually a bronze or copper color. We get red sunsets for the same reason.

This dependence on the atmosphere adds interest to eclipses; if the region through which the sunlight has to pass is unusually thick with cloud, the eclipse will be darker than average, while a transparent atmosphere will give a bright eclipse. Atmospheric conditions cannot be forecast with much accuracy, and there is always the chance of something unusual happening. For example, the two lunar eclipses of 1964 were unusually dark, caused by dust from the huge explosion on the island of Bali in the previous year. Another volcanic eruption, in Mexico, resulted in the dark eclipse of July 6, 1982.

Occultations

The Moon's orbital motion carries it once round the sky in its synodic period of $29\frac{1}{2}$ days, which works out at an hourly velocity of about $\frac{1}{2}°$—its own apparent diameter. This means that it must inevitably pass across and block out the stars in its path, a phenomenon known as an "occultation."

Occultations of fairly bright stars are well worth watching, and can be seen just as well in binoculars as with a large telescope. First of all there is the slow approach of the Moon's limb; bright or dark, depending on the phase. Since it moves among the stars from west to east, disappearance will take place at the dark limb before Full and at the bright limb after. Reappearance, of course, is the reverse. Occultations occurring at the dark limb are especially spectacular, for the lunar disk is naturally invisible against the sky, and the star disappears or flashes out in the fraction of a second. This suddenness forms an incidental proof that the Moon's atmosphere is virtually nonexistent, for a slight trace of air would produce a flickering and fading some time before the true disappearance.

Table 2 *Forthcoming Lunar Eclipses*

DATE	TIME (UT)	MAGNITUDE	DURATION (IF TOTAL) (MIN.)	VISIBILITY UK	VISIBILITY USA
1985 May 4	19·57	1·23	70	Partly	No
Oct 28	17·43	1·08	42	Partly	No
1986 Apr 24	12·44	1·20	68	No	No
Oct 17	19·19	1·27	74	Partly	No
1987 Oct 7	03·59	0·01	Partial	Yes	Yes
1988 Aug 27	11·04	0·30	Partial	No	Partly
1989 Feb 20	15·36	1·28	76	Yes	No
Aug 17	03·09	1·60	98	No	Partly
1990 Feb 9	19·12	1·09	46	Yes	No
Aug 6	14·13	0·68	Partial	No	No

The *Observer's Handbook* lists all favorable occultations, but it must be remembered that the glare of the Moon will prevent all but the brightest stars from being observable near the lunar glare. However, occultations of bright stars, such as Aldebaran, in Taurus, and Spica, in Virgo, are worth looking out for, particularly if the Moon happens to be crescentic rather than gibbous.

Occultations of a given star do not occur regularly each lunation. The apparent path of the Moon around the celestial sphere varies in an eighteen-year cycle; if the path lies in front of a star, an occultation must occur, but for long intervals the Moon will pass clear of the star altogether. We therefore have "seasons" during which occultations of a given object can occur. The current and forthcoming seasons for the four brightest occultable stars in the sky are given below, to the nearest tenth (decimal point) of a year.

Aldebaran (α Tauri): 1977.7–1981.4; 1996.4–2000.0
Spica (α Virginis): 1987.1–1988.6; 1993.9–1995.4
Antares (α Scorpii): 1986.3–1991.4
Regulus (α Leonis): 1979.7–1981.1; 1988.4–1989.9; 1998.3–1999.7

When a planet is occulted, binoculars will show a few seconds' fading prior to complete immersion. This is because a planet, unlike a star, shows a perceptible disk, and it takes the Moon some time to cover it completely.

☆ 4 ☆

The Planets

Most astronomical books are written in the wrong order, and this is no exception. If there is any excuse for starting off with the solar system and working up to the infinitely vaster stellar universe, it is that our nearby surroundings are of much greater personal importance. They are important because of their complete segregation from the rest of the universe, and to see how isolated our planetary system really is, the best way is to watch the course of a single beam of light after it leaves the Sun, traveling at its universal velocity of 186,000 miles per second.

After slightly over 3 minutes it passes Mercury; 6 minutes takes it to Venus; 8 minutes 20 seconds brings it to the Earth, with a $1\frac{1}{2}$-second jump to the Moon. Then comes a big gap; Jupiter, the first of the giant planets, is nearly three-quarters of an hour away, while outermost Pluto requires a journey of $5\frac{1}{2}$ hours before it is reached.

But after passing Pluto's orbit the light-beam can race on and on without the slightest chance of colliding with anything else—for 4 years and 4 months, the distance of the nearest star!

Put in its proper perspective, therefore, the solar system is a compact group. At its center, vastly larger and more important than its family of planets, is our own private star: the Sun. Revolving round the Sun are the nine major planets, ranging from boiling

Mercury to frozen Pluto, with the Earth third in distance.

The whirling Earth is our platform, and because of its position it automatically divides the solar system into two very unequal parts. Inside its orbit revolve the "inferior planets" (Mercury and Venus), while the remainder, the "superior planets" (Mars, Jupiter, Saturn, Uranus, Neptune and Pluto) lie outside.

The inferior planets

Because they are inside the Earth's orbit, Mercury and Venus can never stray very far from the Sun; Fig. 7 shows their movements. When at A an inferior planet is at "eastern elongation," so called because it is then at its greatest angular distance from the Sun, to the eastern side. Gradually its orbital motion moves it in to B, known as "inferior conjunction." At this position it is closest to the Earth, but it is too near the Sun to be observable; in any case the night side is turned towards us, like the Moon when New. It then moves out to C, which is western elongation, and finally it closes in towards the Sun again at D. At this position, "superior conjunction," it is at its most distant. The times taken to complete these paths are approximately 116 days for Mercury, 584 days for Venus. These times take into account the Earth's own rotation around the Sun, and are the planets' synodic periods; a planet's

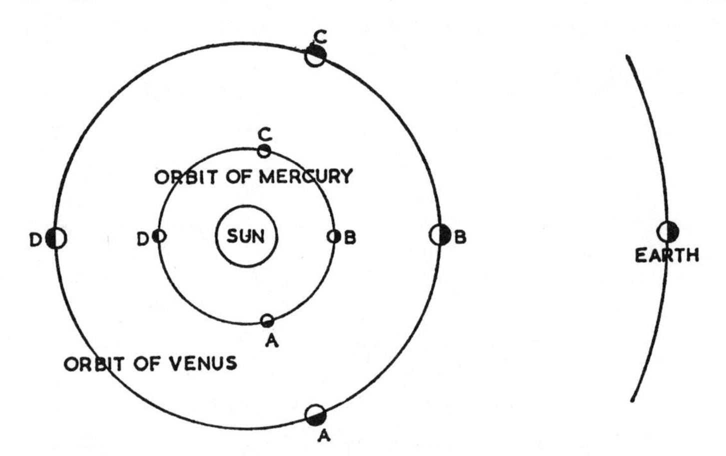

Fig. 7. *Movements of the inferior planets* (not to scale)

true revolution period, or year (88 days for Mercury, 225 days for Venus) is called the sidereal period.

We see these orbits very inclined because all the planets move in almost the same plane; if we drew a plan of the solar system on a sheet of paper it would not be very far from the truth. Unfortunately, with the exception of Pluto (which seems to be exceptional in most ways) Mercury and Venus are the worst offenders. Mercury's orbit is tilted at an angle of 7° to our own, Venus's at $3\frac{1}{2}$°, and this has the unfortunate effect of making transits across the Sun at inferior conjunction rather rare; they usually pass either above or below the solar disk.

The best chance of seeing an inferior planet is obviously when it is at its greatest angular distance from the Sun, near either A or C. Western elongations must be seen before sunrise, while at eastern elongation Mercury can set up to two hours after the Sun, and Venus, which moves in a larger orbit, can set four hours after and so be visible against a really dark sky. Not all elongations are, however, equally favorable—something which comes about through the $23\frac{1}{2}$° tilt of the Earth's axis (Fig. 8).

Position A shows the Earth at the time of northern midsummer, when the north pole is turned its greatest extent towards the Sun —at this time it appears directly overhead at noon to someone in latitude $23\frac{1}{2}$°. Three months later it is at B, the autumnal equinox (September 23), when the Sun is in the plane of the equator and appears to be traveling south. At C (northern midwinter) it is

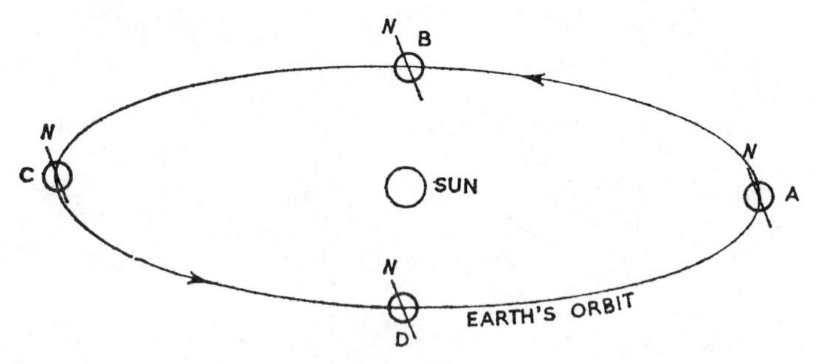

Fig. 8. *The seasons* (not to scale)

This shows the Earth's orbit as seen from a point out in space somewhat north of the orbital plane. The Earth's axis remains fixed in direction.

$23\frac{1}{2}°$ south; after that it starts to drift north again and by March 21 (D) it is on the equator once more, traveling north, at the vernal equinox.

This movement of the Sun can be represented as in Fig. 9, which is a panoramic view of the sky in the region of the celestial equator. The wavy line represents the Sun's apparent path north and south of the equator throughout the year, and it is known as the "ecliptic."

Since the Sun lies exactly in the plane of the Earth's orbit, the ecliptic is nothing more than the projection of our orbit in the sky. This is most easily understood by supposing the Sun to move round the Earth; it lies in the plane of our orbit, and wherever it moves it marks out that plane. Suppose for a moment that the Earth's axis were vertical. The equator and the ecliptic would then coincide, the Sun would always have the same altitude, and there would therefore be no seasons.

All the planetary orbits are in very nearly the same plane, and this means that their paths must follow the ecliptic very closely. Just as the Sun is highest (to a northern observer) when crossing the midsummer point on June 21, so is a planet. When it is near the midwinter part of the ecliptic it never rises high in the sky and so is difficult to observe adequately.

We can now return to the inferior planets, which present a special problem: the prime consideration here is their altitude above the horizon at sunset or sunrise. Fig. 10 shows two eastern elongations at sunset: one on March 21, the other on September

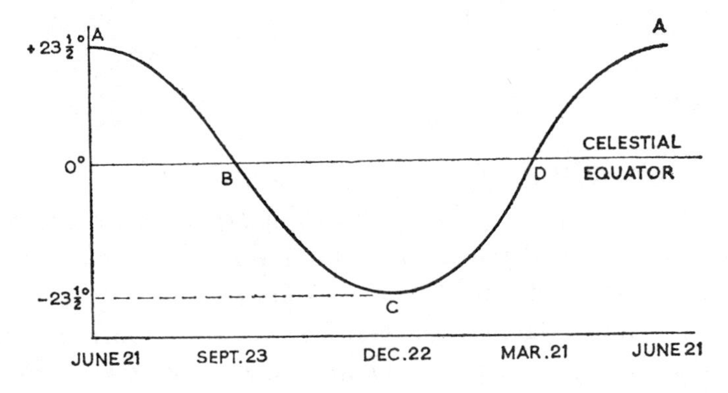

Fig. 9. *The ecliptic*

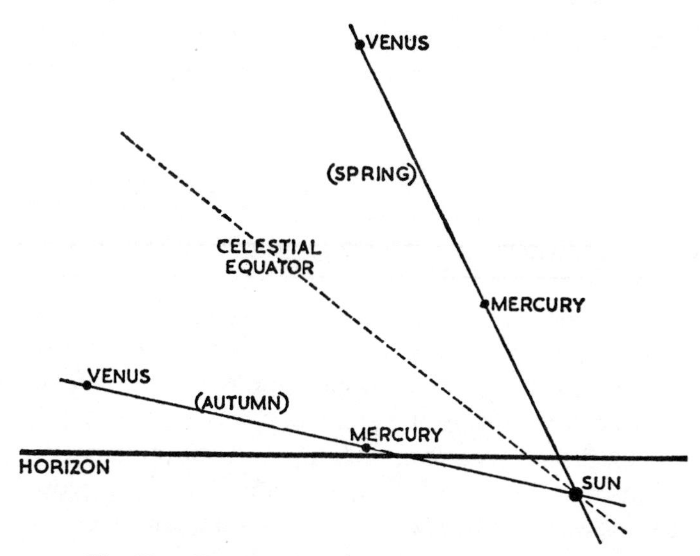

Fig. 10. *Evening elongations in spring and autumn*

Note that the planet's direction of setting is parallel to the celestial equator, not to the ecliptic.

21. In both cases the angular distance from the Sun is the same, but the spring sunset is obviously far more favorable; the planet is in the north-bound part of the ecliptic, while in autumn it is in the south-bound region. Evening elongations are therefore best seen in the spring, and a little thought will show that the autumn is the best time to observe a western, or morning, apparition. These circumstances apply to both the inferior planets.

MERCURY

Mercury is a challenge for binoculars. Its greatest possible elongation from the Sun is only 28°, which must obviously occur when it is at aphelion, and because its orbit is very eccentric a perihelic elongation carries it only 18° away from the Sun (Fig. 11). By a piece of cynical planning on Nature's part the more favorable aphelic elongations always occur when the planet is well south of the equator, so that northern observers have to be content with far less spectacular views of the innermost planet. Southern observers are also blessed with the best views of Mars at its most favorable appearances.

From latitudes of 50° or more, Mercury is an elusive object in the sense that it will probably never be noticed by accident, although it is easily spotted at a favorable elongation. Having established the favorable week or ten days during which it is best placed, search the twilight sky about $\frac{3}{4}$ hour from sunrise or sunset. The planet will appear as a bright binocular object some 11° above the horizon and about 9° southwards of the center of the twilight glow.

This assumes an elongation of 20° from the Sun. Once located, there should be no difficulty in seeing it with the unaided eye. From southern States its altitude will be somewhat greater, and, setting later, it can be followed into a darker sky. Given the essentials of a low horizon and clear weather, the innermost planet is much easier to find than most people seem to imagine.

In the British Isles, I have occasionally been able to observe Mercury on several successive days at a favorable elongation. However, during a sojourn on the island of Crete (latitude $34\frac{1}{2}$° north), the planet was much more easily seen because of the steeper angle of the ecliptic with the horizon, and the consequently greater altitude of the planet at twilight. Indeed, it was once observed on twenty-four consecutive evenings.

Mercury occasionally passes close to a bright star. These so-

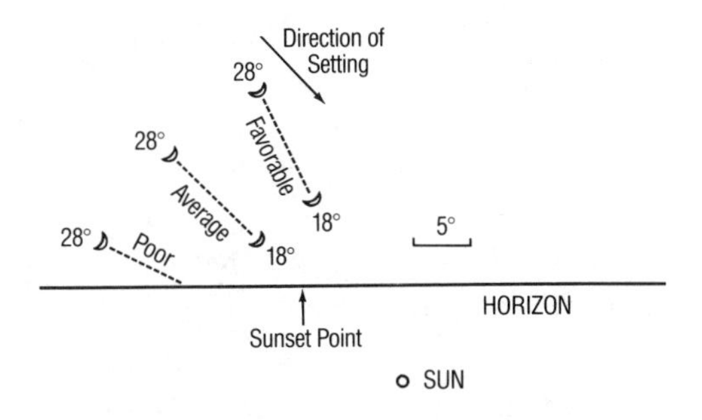

Fig. 11. *Perihelic and aphelic elongations of Mercury*

The diagram represents the different positions that Mercury can adopt in the evening sky, some 45 minutes after sunset, as seen from a latitude of 45° N.

Table 3 *Forthcoming Elongations of Mercury*

The angular distance from the Sun at greatest elongation is given. As a general rule, evening elongations are best observed in the spring, and morning elongations in the autumn.

	EASTERN (EVENING)	WESTERN (MORNING)
1983		October 1 (18°)
	December 13 (21°)	
1984		January 22 (24°)
	April 3 (19°)	May 19 (26°)
	July 31 (27°)	September 14 (18°)
	November 25 (22°)	
1985		January 3 (23°)
	March 17 (18°)	May 1 (27°)
	July 14 (26½°)	Aug 28 (18°)
	November 8 (23°)	December 17 (21°)
1986	February 28 (18°)	April 13 (28°)
	June 25 (25°)	August 11 (19°)
	October 21 (24½°)	November 30 (20°)
1987	February 12 (18°)	March 26 (28°)
	June 7 (24°)	July 25 (20°)
	October 4 (26°)	November 13 (19°)
1988	January 26 (18½°)	March 8 (27°)
	May 19 (22°)	July 6 (21°)
	September 15 (27°)	October 26 (18°)
1989	January 9 (19°)	February 18 (26°)
	May 1 (21°)	June 18 (23°)
	August 29 (27°)	October 10 (18°)
	December 23 (20°)	
1990		February 1 (25°)
	April 13 (20°)	May 31 (25°)
	August 11 (27°)	September 24 (18°)
	December 6 (21°)	

called appulses can be interesting, but they are usually difficult to observe; Mercury, shy as it is, is really much brighter than any of the stars it can approach. Appulses to Venus are more favorable, for Venus is bright enough to be seen under almost any conditions and can serve as a guide to the fainter planet.

Apart from the satisfaction of glimpsing it in the twilight sky, Mercury's main interest lies in its occasional transits across the

Sun. The last was on November 9, 1973, when it could easily be seen with protected binoculars as a black speck moving slowly across the solar disk. Transits usually take about five hours, and the next will not occur until November 13, 1986.

VENUS

Venus is anything but modest. Every now and then the morning or evening sky contains a brilliant object that puts even the brightest stars to shame; when very near the horizon it shines like a distant lamp. It can sometimes be seen in broad daylight, and at a favorable elongation; when it glares down from a dark sky, it casts a distinct shadow.

There are several reasons why Venus is such a startlingly different proposition from Mercury. It is much larger—7,700 miles in diameter as against 3,100—and instead of being an airless, barren globe it is covered with dense cloud which reflects nearly 60 percent of the sunlight falling on it. Add to this its much wider elongations (they average 47°, hardly varying at all because the orbit is almost circular), and it is hardly surprising that Venus is the most brilliant planet in the sky. In addition its minimum distance from the Earth is a mere 26 million miles, as against 35 million in the case of the second-nearest neighbor, Mars.

To make the search for the Planet of Love worthwhile we have to manufacture difficulties. One is to look for it during the day; another is to catch it near superior or inferior conjunction. The best game is to follow it in towards inferior conjunction at the end of an evening elongation, to see how near the actual day of conjunction it can be spotted. At the very favorable conjunction of April 6, 1977, Venus was seen by a number of observers on the very morning of its passage past the Sun, appearing as a naked-eye object low in the dawn glow.

The phases of Venus

Fig. 12 explains the phases of Venus (or Mercury). In the first position, superior conjunction, its illuminated hemisphere is turned fully towards us and it appears as a circular disk, shrunken by distance. Gradually its orbital motion sweeps it out east of the Sun and it comes into the evening sky, its phase lessening but its brightness increasing through its steady approach towards the Earth. At elongation it appears as a perfect half, and as it curves in toward inferior conjunction the disk quickly expands and the phase narrows into a crescent. At a certain point—when 28% of the disk is illuminated—maximum briliancy occurs, distance and phase combining to the best advantage. When it reappears in the morning sky the phases are run through in the reverse order.

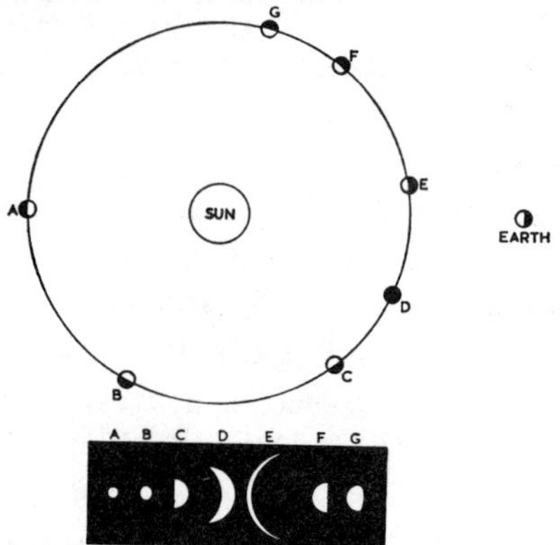

Fig. 12. *The phases of Venus* (not to scale)

Good quality binoculars should show the crescent phase quite easily, and the thing to remember is that magnification is not nearly so important as optical quality. This is so in all branches of astronomy, but Venus is an exceptionally severe test; its briliance is so intense that it produces false glare in even a good telescope, and this glare masks the outline. However, in twilight or full daylight, a pair of 8 × 30 binoculars will show the crescent phase very clearly, and I have also seen it well with 6 × 24 glasses.

The *Observer's Handbook* gives all necessary position and phase data, and a table of future elongations is given below.

Daylight observation

The first step in finding Venus during the day, supposing that it is too close to the Sun to be found by simple sweeping of the general area, is to locate its position with reference to that of the Sun. Suppose that Venus is an evening star, following the Sun across the sky. At around the time of local noon (say, within half an hour of midday) set the binoculars on stand and point them to the Sun. Its image can be caught on a sheet or card held a few inches behind the eyepieces. They are then clamped firmly and left for a time interval equal to the difference of right ascension (see page 84) between Venus and the Sun, an amount to be found in any astronomical almanac. After this interval has elapsed, Venus must have the same southerly azimuth as that of the Sun when the binoculars were set up. After swinging the binoculars vertically, either up or down, through the angular difference of their declina-

Table 4 *Forthcoming Movements of Venus*

Date	Event
1983 Nov 4	Morning elongation
1984 Jun 15	Superior conjunction
1985 Jan 22	Evening elongation
Apr 3	Inferior conjunction
Jun 13	Morning elongation
1986 Jan 20	Superior conjunction
Aug 27	Evening elongation
Nov 5	Inferior conjunction
1987 Jan 15	Morning elongation
Aug 23	Superior conjunction
1988 Apr 3	Evening elongation
Jun 12	Inferior conjunction
Aug 22	Morning elongation
1989 Apr 5	Superior conjunction
Nov 8	Evening elongation
1990 Jan 18	Inferior conjunction
Mar 30	Morning elongation
Nov 2	Superior conjunction

tions (see page 84), the planet should be found near the center of the field of view.

To take an example, suppose that the crescent planet was being sought on January 13, 1974. Its elongation from the Sun on this date was only 17°, and the 4 percent crescent, seen through the low skies of a northern winter, would be difficult to locate without knowing exactly where to look. From an almanac, the position of the Sun on that date is found to be R.A. 19 hours 39 minutes, Dec. 21° 29′ south of the celestial equator. The corresponding position of Venus is at R.A. 20 hours 43 minutes, declination 13° 49′ south, so it is higher in the sky than the Sun. Having pointed the binoculars sunward at around the time of local noon (a half hour's difference either way will be immaterial), an interval of 1 hour 4 minutes would bring Venus to the same meridian. Elevating the binoculars by 7° 40′ (a protractor fixed to the stand would be helpful) should complete the search.

This "drift" method cannot be used when Venus is a morning star, preceding the Sun across the sky. Instead, its approximate position must be found by direct offsetting from the Sun. Remembering that 1 hour of right ascension is equivalent to about 15°, the planet could, in the above example, have been found by moving the binoculars 8° above the Sun and 16° east. These angles can, with experience, be estimated with sufficient accuracy to bring the planet somewhere in or near the binocular field, so that a moderate amount of careful sweeping will locate it. A useful rule to remember is that the angle subtended by an outstretched hand, with thumb and little finger extended, is about 20°. The method is reliable only when the Sun and Venus are near the meridian, so that their apparent east-west drift across the sky is approximately parallel to the horizon.

Although binoculars will readily show Venus in a really clear daylight sky, it must be remembered that the slightest whitish haze (permanent over most cities) can make it invisible even using a small telescope. Under good conditions it is a fairly easy naked-eye object when near maximum brilliancy, once its exact position is known. Similarly, Jupiter and even Mars have been recorded as naked-eye daylight objects.*

*It is surprising what can be seen from the most unexotic site. Some time ago, when Venus and Jupiter were near each other in the afternoon sky, an observer picked up both objects and showed them to his nonastronomical colleagues. These naked-eye observa-

Appulses to bright stars occur in the same way as with Mercury, but they are usually much more spectacular; when they happen in a bright sky, Venus can be used as a guide to find the star. The best example of this was on July 7, 1959, when Venus actually occulted Regulus not long after noon. Weather was generally good at the time, and the star could be seen with 3- and 4-inch telescopes once Venus was found.

Occultations of Venus by the young or old Moon are not infrequent, although any given observer may not see one for some years. Regularly, however, we are treated to the sight of the brilliant evening or morning star hanging in the twilight sky next to a lunar crescent, with perhaps Mercury or Jupiter in attendance; and such conjunctions need no optical aid at all to enhance their beauty.

Transits of Venus across the Sun are very rare. They occur in pairs, the last being in 1874 and 1882, and the next are not due until June 7, 2004, and June 5, 2012.

The superior planets

A marked change comes over planetary movement when we shift outside our own orbit. Our gaze is not necessarily in the vicinity of the Sun, for it is obvious from Fig. 13 that a superior planet travels independently right round the ecliptic.

The planet P, in the diagram, is closest to us at position P_1 when the Earth is at E_1—it is opposite the Sun in the sky, like the Full Moon, and the position is therefore known as "opposition." A little thought shows that at opposition a planet rises at sunset, is due south at midnight, and sets at sunrise; in other words, it is above the horizon all night.

Opposition passes and planet P moves along its orbit. But the Earth is also moving, and because it is closer to the Sun it travels faster. Thus, a year later it has returned to E_1, while the planet has only reached P_2. So all the Earth has to do is continue on its way for a little longer before it catches up with the planet and forms another opposition, at E_2 and P_3. The time between successive

tions were made through a window from a factory in the industrial town of Luton, England!

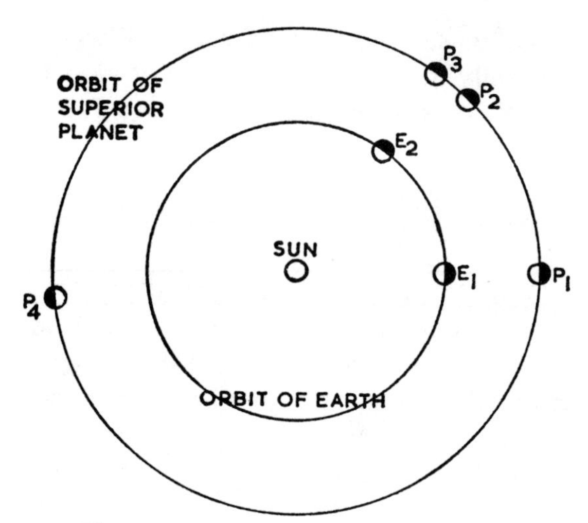

Fig. 13. *Movements of a superior planet*

The orbits are drawn to scale. Notice how the more distant planets not only move more slowly but also have a much greater distance to travel. Pluto will be at perihelion in 1989.

oppositions is the planet's synodic period. The intermediate position, P_4, when the planet is very near the Sun (and may be directly in line with it), is called "conjunction."

The more distant the planet the shorter is its synodic period. Suppose that there was a very remote planet whose yearly motion was negligible; then all the Earth would have to do would be to return to E_1 for another opposition. The planet's synodic period would be just one year, and it would never seem to move along the ecliptic.

Remote Pluto takes 248 years to circle once. This means that when the Earth has returned to the starting point after an opposition, it needs to go only about $\frac{1}{248}$ of a circuit further to catch Pluto up again—in other words, about $1\frac{1}{2}$ days. Jupiter, much closer to the Sun, has a sidereal period of about 12 years, so that oppositions take place $\frac{1}{12}$ of a year later each year, or at intervals of about 13 months.

Mars, the closest of the superior planets, is a rather more special case. Its sidereal period is just under two years, which means that by the time the Earth has returned to E_1 it has completed half an orbit, and is on the far side of the Sun, near conjunction. In this position, due to its greatly increased distance, it is unobservable, and in any case it is in too bright a sky to be visible. It takes another orbit and a couple of months before the three line up again, and Mars therefore has the longest synodic period of all: approximately 780 days.

At opposition a superior planet is opposite the Sun in the sky, so it follows that oppositions occurring when the Sun is in the lowest part of the ecliptic (i.e., during the winter) are the most favorable, so far as altitude is concerned. At a summer opposition a planet hugs the horizon and is badly placed for observation. Other considerations also apply to Mars, the first planet we meet on the journey beyond the Earth's orbit.

MARS

The orbit of Mars is noticeably eccentric (Fig. 14). Its mean distance from the Sun is 142 million miles, but this varies from 129 million at perihelion to 154 million at aphelion. Neglecting the variation in the Earth's distance from the Sun, opposition distances can therefore vary from 35 million miles (when it occurs with Mars at perihelion) to over 60 million at aphelion.

The last perihelic opposition occurred in August 1971, when to northern observers Mars shone redly near the southern horizon, appearing, for a few weeks, more brilliant than Jupiter. The cycle of ever-increasing opposition distances has now started; in 1973 it never approached closer than 40 million miles, and conditions continued to deteriorate until the last aphelic opposition, in February 1980, when it appeared fainter than the star Sirius. Not until the oppositions of 1986 and 1988 will Mars be favorably situated for observation.

From the point of view of northern observers perihelic oppositions are not necessarily the best. We saw earlier that Mercury is more favorably placed for southern observers at its widest elongations, and they also have the best of Mars; at perihelion it is always well south of the equator. In 1971, for instance, it was in declination—20°, which meant that from a northern latitude of 45° it was only 25° above the horizon at its highest point. In 1973, on the

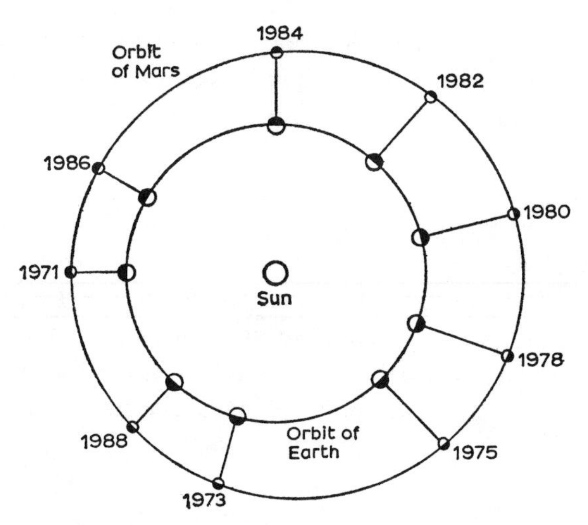

Fig. 14. *Oppositions of Mars, 1971–1988*

The two orbits are drawn to scale.

other hand, its declination was +10°, giving an altitude of 55°. Atmospheric conditions are so much better at a high altitude that northern observers had a better view, despite the smaller apparent size.

Tracking the motion of Mars

Mars's disk is nearly always too small to look like anything but a star in normal binoculars, though at a good opposition it is just about distinguishable. However, they can do interesting work in tracking its path among the stars (the word "among" of course, should not be taken too literally!).

Norton's Atlas can be used as a basis for this work, but unfortunately it shows only naked-eye stars; and a more comprehensive atlas may not be available. The best and most interesting procedure is to make a homemade chart, marking out the region through which Mars will pass over a suitable interval (say, two months). Copy the *Norton's* stars on a large sheet of paper, multiplying the scale several times, and fill in the fainter ones by observation. As well as giving good practice in drawing star-fields, it is interesting to later compare the chart with a more comprehensive map to get an idea of its accuracy. Draw in lines of R.A. and Dec., so that the observed and predicted positions of Mars can be compared. Right ascension and declination, the longitude and latitude of the celestial sphere, are explained on page 84.

The weekly motion of Mars along the ecliptic is obvious even with the naked eye, and binoculars will show changes over a much smaller interval. As the positions are plotted join them into the most natural-looking curve that will fit, and date each entry.

Retrograde motion

If you follow a planet night after night, the textbooks say, you will eventually notice a most extraordinary piece of behavior. Its normal easterly progress will slow down and slip into westerly backtracking; after a time easterly motion will resume. The much easier way of discovering this "retrograde" motion, by glancing at a planetary ephemeris, is less romantic but decidedly timesaving.

All the planets exhibit retrograde motion—even Mercury and Venus, which, however, camouflage it in the starless twilight sky. Fig. 15 explains it for a superior planet, when it occurs through the Earth's greater orbital velocity.

From our vantage point we see the planet's motion projected against the sky background—not the three-dimensional motion that it really is—and this apparent movement is called "proper motion." Starting with the Earth at *A,* we are looking at the planet more or less along the line of our own orbital travel, which means that it has no effect on the planet's proper motion; if we are heading directly towards an object it does not appear to shift relative to the background, though of course it expands in size.

By the time the Earth reaches *B* it is no longer traveling directly towards the planet, and its motion is having some effect; this effect is to slow down the planet's proper motion, since both the velocities are in the same general direction. Subsequently, as we approach opposition, the planet moves more and more slowly until the Earth reaches *C,* when the two velo-

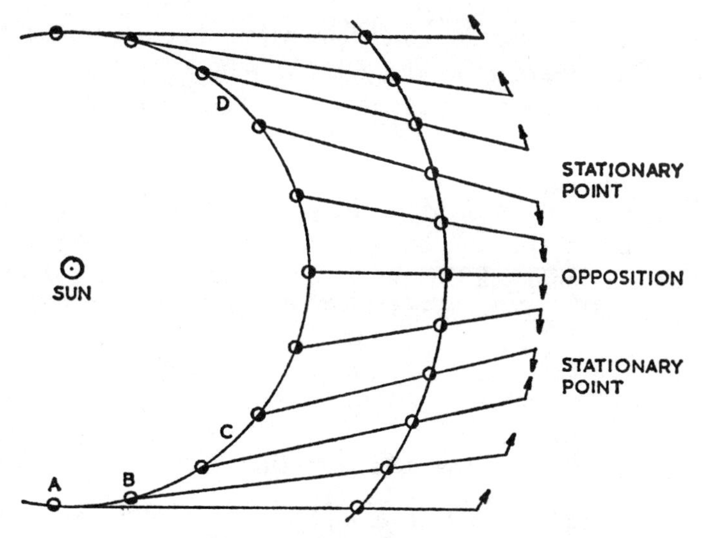

Fig. 15. *Retrograde motion*

The arrows indicate the apparent or proper motion of the planet.

cities are effectively matched. After that the Earth's greater speed has the upper hand; it sweeps past the planet until *D,* when the velocities match again, and after that things proceed normally. The result is that the planet appears to backtrack for a time around opposition.

Retrograde motion was one of the chief headaches of Ptolemaic enthusiasts; an Earth-centered solar system could not account for it. It was eventually explained, together with other discrepancies, by supposing each planet to move in a small circle, the center of which moved round the Earth in a larger circle. These "epicycles" had a remarkably high birth rate as more and more inaccuracies appeared; Ptolemy himself provided his system with over thirty, and later observers were forced to add more.

Fig. 16 shows the loop Mars performed in the sky during the opposition of 1975–6. Because of the angle at which their orbits are inclined, Mars, Jupiter and Pluto exhibit fairly open loops, while Saturn, Uranus and Neptune seem to retrograde along their own paths.

Mars poses its own problems, but these are strictly beyond the range of binoculars. So are its two satellites, neither of which is

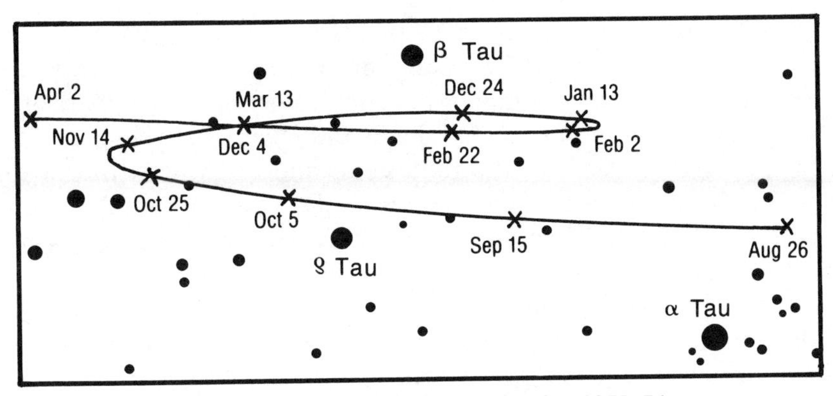

Fig. 16. *The path of Mars in the sky, 1975–76*

The crosses show the position of the planet at 20-day intervals during its passage through Taurus. Opposition occurred on December 15. The area covered by the map is about 15° by 33°.

more than ten miles across; it takes a fair-sized telescope to reveal them at all.

Changes in brightness

The brightness of Mars changes very considerably throughout an apparition. The main factor is its changing distance from the Earth, but there is also its elliptical orbit to be taken into consideration; near perihelion it experiences markedly stronger illumination. There is also a slight phase effect when it makes the greatest angle between the Earth and the Sun, and telescopically it can sometimes appear like the Moon two or three days from Full.

When dealing with the average variable star (page 78) things are reasonably straightforward; it is simply compared with nearby stars whose brightness or "magnitude" are known. Star-fields do not change, and the same stars are always available for comparison.

But a planet, particularly a fast-moving and bright planet like Mars, presents two difficulties. First of all it wanders along the ecliptic, so that a star which is conveniently close in June may be nowhere to be seen by December. Another rather more serious objection is that the number of suitably bright stars is very limited; at its brightest Mars shines much more vividly than any star, and even at other times its marked orange color makes comparison with normal white stars both difficult and uncertain. There are, in fact, only three stars of the right color and brightness; they are α Boötis (Arcturus), mag. −0·06; α Tauri (Aldebaran), mag. 0·86, and α Scorpii (Antares), mag. 0·89. An explanation of magnitude classification is given on page 83.

These comparison stars will enable estimates to be made during the fainter stages of the apparition, but Mars far outshines even Arcturus at the worst opposition.

The rotation of Mars

One claim which has frequently been made is that the rotation of Mars (its day is 24 hours 37½ minutes long) can be detected with nothing stronger than the naked eye—simply by its color. The best analogy is the Earth itself. To an observer on Venus (supposing he could see anything through the clouds) the Earth's axial spin would present alternately the vast spread of the Pacific, then the main land-mass—two completely different hues; the change from the brown of the continents to the blue of the ocean would be obvious. But Mars, though similar in having most of its extensive dark areas confined to one hemisphere, has much less contrast between these patches and the ocher desert.

The challenge is there, however, and it is certainly worth investigating by comparing its hue with nearby reddish stars. Binoculars are useless for this work which, like magnitude determination, requires a very large field of view; in any case, the brightness of Mars needs no augmenting.

It certainly needed no telescope to detect the unusual Martian conditions during the very favorable 1956 opposition. Instead of appearing pale orange its color was almost yellow. Extensive dust storms provided the cause; they obscured the surface features so effectively that things did not revert to normal until 1960. Similar storm activity during the 1971 apparition also made the planet look noticeably pale to the naked eye.

A table of future Martian movements follows.

THE MINOR PLANETS

On the first day of January 1801, an Italian astronomer came across a starlike object that showed a planet's motion. This, the first minor planet to be discovered, was the result of much devoted searching, and fittingly enough is the largest. It is 600 miles across, named Ceres, and caused a stir in the astronomical world quite unmerited on the grounds of size alone; but it raised expectancy of many sisters, and even the most optimistic hopes have been overfulfilled.

For Ceres was merely the prelude to the subsequent cascade of minor planet discoveries. Over two thousand have now been detected, most of which are extremely faint; but some of the brighter ones, which were naturally enough among the first to be discovered, are well within the range of binoculars. The *Observer's Handbook* lists the positions around opposition of several of the brighter minor planets.

The way to identify a minor planet is to draw all the stars in the field known to contain it. When the sketch is compared with the view three or four nights later, one of the "stars" will have moved, betraying its planetary nature. Once it is identified, a minor planet can be followed until the evening twilight swallows it up.

Vesta can sometimes just be seen with the naked eye, but most of the time it is well below the threshold, and the others are permanently invisible without optical aid. There has recently been some revival of interest in magnitude observation of these bodies, and binocular work by several enthusiastic observers has revealed significant inaccuracies in the brightness predictions issued by professional observatories.

Table 5 *Forthcoming Oppositions of Mars*

DATE	CONSTELLATION	MAG.
1984 May 11	Libra	−1·7
1986 Jul 10	Sagittarius	−2·4
1988 Sep 28	Pisces	−2·6
1990 Nov 27	Taurus	−2·0

JUPITER

The giant of the solar system never approaches closer than about 370 million miles (ten times the minimum distance of Mars), but its huge disk, 88,700 miles across, makes it appear much larger than the Red Planet. Binoculars easily reveal it to be something other than a star, and with a little perseverance they should show its four brightest satellites.

Jupiter circles the ecliptic much more slowly than Mars; it takes nearly twelve years to complete a circuit, and because of this there is less motion to track. The visibility chase, however, can be continued longer, and Fig. 17 explains why.

A planet is at its minimum distance at opposition, and maximum distance at conjunction; in the case of Mars the average distances are 48 million miles (M_1) and 234 million miles (M_2), meaning a fivefold change. Jupiter, however, is moving in a much larger orbit. Its mean solar distance is 483 million miles; at mean opposition it approaches to within 390 million miles (J_1), while at conjunction it recedes to 576 million miles (J_2)—only $1\frac{1}{2}$ times the opposition value. It is obvious that the more remote the planet, the closer the two values will match.

Planetary brightness follows the normal inverse-square law; at twice the distance a planet is four times as faint. It therefore follows that Mars at conjunction is only $\frac{1}{25}$ as bright as at opposition, while Jupiter loses only half its light. Bearing in mind the additional fact that Jupiter nearly always outshines Mars at opposition, it becomes obvious why the giant planet is much better prey for binoculars in the twilight evening sky as it moves in towards the Sun.

The satellites of Jupiter

The four bright satellites provide interesting entertainment. If they were removed from Jupiter's presence and scattered in the sky they would be visible with the naked eye; what makes their detection more difficult is the close proximity of the brilliant planet. The best time to look at them is when they are at their individual elongations.

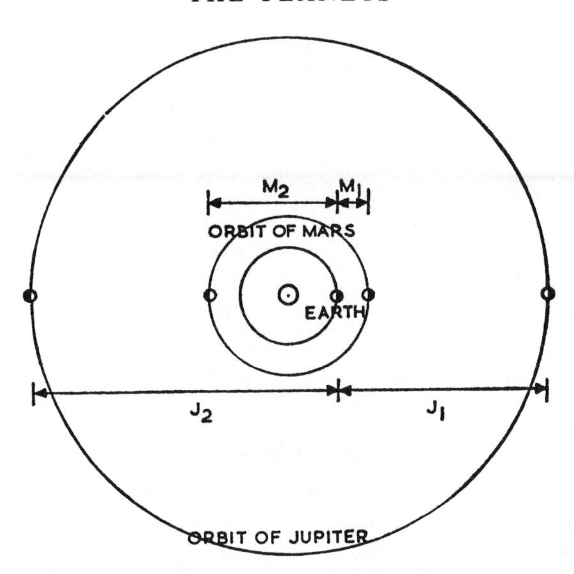

Fig. 17. *Relative brightness of Mars and Jupiter*

The orbits are drawn to scale.

The closest, Io, is naturally the most difficult; it is never more than three planet-diameters away. Europa is slightly more cooperative, but the third, Ganymede, is easy—it has frequently been seen with the naked eye. The outermost, Callisto, wanders up to 12′ away from Jupiter, but it is the faintest of the four.

These satellites, discovered by Galileo in 1610, are always spread out in a straight line; they move exactly in the plane of Jupiter's equator, and since its pole is nearly vertical (as against our $23\frac{1}{2}°$ tilt) we always see their orbits edge-on. Jupiter's four moons, the brightest of fourteen, provide one of the spectacles of the sky—a sight not denied the most modest optical aid.

SATURN

It is a sad piece of irony that Saturn, the most beautiful object in the entire sky, should be so unremarkable without a powerful telescope. It is sufficiently bright to be easily identifiable, so there is no interest there; its movement is too slow to be of continuous

Table 6 *Forthcoming Oppositions of Jupiter*

DATE	CONSTELLATION	MAG.
1984 June 29	Sagittarius	–2·2
1985 August 4	Capricornus	–2·3
1986 September 10	Aquarius/Pisces	–2·4
1987 October 18	Pisces	–2·5
1988 November 23	Taurus	–2·4
1989 December 27	Gemini	–2·3
1991 January 28	Cancer	–2·2

interest, and it has no bright satellites to match the easy visibility of Jupiter's retinue. However, the largest of its ten moons, Titan, which is larger than the planet Mercury, is detectable with 8 × 30 binoculars as a faint, starlike object that takes 16 days to achieve one revolution around its primary.

If they are of good quality, high-powered binoculars will show Saturn's elliptical outline quite distinctly (the general form of the planet was discovered by Galileo with a telescope magnifying only × 30, and its optical quality was very inferior by present-day standards). However, this whole matter raises a question of fundamental importance which is certainly worth going into in a little more detail—the question of "resolving power."

The position can be illustrated as follows. Suppose we point a pair of 20 × 60 binoculars at Saturn and focus carefully; at its best, the image will appear as a tiny elliptical blob of light, with no detail visible at all. Now use the same magnifying power on a small astronomical telescope of 4 or 5 inches aperture. The image will naturally be the same size, but it will show far more detail; the ball of the planet can be distinguished from its ring system, and may show fragmentary markings. Not very well, of course, but the superiority of the view will be undeniable. It is due to the improved resolving power of the larger telescope.

This matter of resolving power is of great importance in astronomical work, and it explains why big telescopes can detect much finer lunar or planetary detail than smaller ones; the secret is not so much magnification as aperture. A 2-inch telescope, for instance, could not detect a lunar cleft less than 400 yards across, no matter what magnification was used, whereas a 9-inch telescope could show clefts down to 100 yards across. Resolving

Table 7 *Forthcoming Oppositions of Saturn*

Saturn's considerable variation in brightness is due to the changing presentation of the ring system. When widest open, at which time their face is 28° from the edge-on position, they reflect more sunlight than the planet itself. Consequently, the planet will appear particularly bright in the late '80s, when the rings are well presented. Maximum presentation occurs in 1988, with the north face visible.

DATE	CONSTELLATION	MAG.
1984 May 3	Libra	0·0
1985 May 15	Libra	–0·1
1985 May 27	Scorpius	–0·2
1987 June 9	Ophiuchus	–0·3
1988 June 20	Sagittarius	–0·3
1989 July 2	Sagittarius	–0·2
1990 July 14	Sagittarius	–0·1

power is an inherent part of the optical system, and within certain limits it is unaffected by magnification.

There is also the question of optical excellence. Despite what advertisements boast, binoculars rarely even approach the perfection of a good astronomical telescope. There is a simple reason for this: it is not necessary. They use a relatively low magnification and are normally confined to terrestrial viewing which is much more tolerant of slight imperfections than a bright star—the severest test any telescope can be given.

THE OUTER PLANETS

Uranus and Neptune, the brightest of the three newly discovered members of the Sun's family, are not exactly ostentatious—but neither are they anything like so difficult to find as most people seem to believe. Uranus is in fact visible with the naked eye when its position is known, and its slow drifting round the zodiac (one circuit takes 84 years) can be followed from season to season just by casual observation. The periodic time of Neptune is 168 years; binoculars are needed to find it, and a few nights must be spent

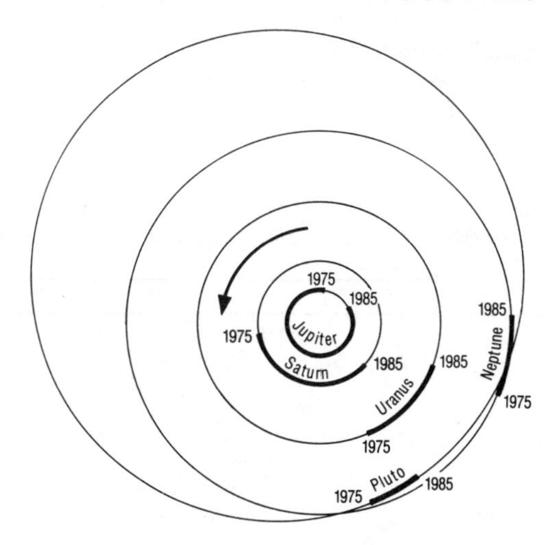

Fig. 18. *Movements of the superior planets, 1975–85*

The orbits are drawn to scale. It should be noticed that the more distant planets not only move more slowly; they also have to cover a much larger orbit. Pluto ceased, temporarily, to be the outermost planet in 1969.

identifying it by its motion, but once it is recognized there is no danger of it slipping away among the stars.

It is a curious paradox that the well-equipped amateur, with a telescope of perhaps 15–30cm aperture, finds himself unable to see more than a small disk representing either of these planets and can do no useful visual work at all; whereas the binocular observer, to whom they appear as no more than starlike points, can make estimates of their brightness that may be of some quantitative use. In 1968, for example, the writer observed both these planets using 7 × 50 binoculars, comparing their magnitude with that of neighboring stars, and discovered that Uranus was significantly brighter than its predicted magnitude, while Neptune was noticeably too faint. These results, made with the most modest of equipment, were incorporated in subsequent editions of the annual *Handbook of the British Astronomical Association*.

There is some slender evidence that the brightness of both these planets varies, by perhaps a few tenths of a magnitude, due possibly to the outbreak of light or dark features on their surfaces, and the binocular observer should attempt to make magnitude estimates (as described on page 84) on three or four occasions during each apparition. At present (1983) Uranus lies in the constellation Ophiuchus, while Neptune is in Scorpius.

Though considerably smaller than Jupiter and Saturn, Uranus and Neptune are still giant-sized compared with the Earth, with diameters of 29,200 miles and 30,100 miles. But distant, lonely Pluto is inferior to everything except Mercury; its remoteness, to the tune of 3,650 million miles, carries it beyond the reach of most amateur telescopes, and its orbital crawl makes it spend about twenty years in each constellation. At present, it is to be found in the constellation Virgo.

In their different ways, then, all the planets except Pluto offer something to the binocular observer. They are the ideal instrument for picking up flighty Mercury, and they can find Venus in daytime and show its crescent phase. Favorably placed, Mars may just show a disk and will certainly reveal its warm tint and its rapid motion in front of the stars. A number of minor planets can be followed along their inconspicuous paths. Jupiter can be found on most nights of the year, and the smallest glass will show at least two or three of its moons, while Saturn shows a nonstellar image and its largest satellite, Titan, may be visible. Finally, the starlike Uranus and Neptune can be watched for possible brightness changes. There will be something for the planetary observer to look at on every clear night.

☆ 5 ☆

Comets, Meteors, Aurorae

There is frequent confusion in the public mind between meteors and comets. Whenever the appearance of a bright comet is predicted, one hears of people having seen it "streak across the sky"; scores of such observations were reported when Comet Kohoutek appeared in 1973–1974. Anything streaking across the sky must be a meteor; cometary bodies move so slowly among the background stars that an hour or two's careful watching may be needed to detect any movement. Comets and meteors do indeed enjoy at least a tenuous relationship, but this is confined to their physical nature rather than their appearance.

A great deal of mystery still surrounds the nature of comets, and the spectacular appearance of a bright visitor occasions general interest and even, in some quarters, unease, although this is a hangover from the astrological age, when a comet was taken to herald disaster. A comet is a diffuse cloud of dust and frozen gases, possibly with a rocky nucleus a few miles across, the whole being bound into an entity some hundreds of miles in diameter under its own weak gravitational attraction. Whether these insubstantial bodies were originally members of the solar system—perhaps the fragments of a distant primordial cloud, beyond Pluto, that never condensed into a planetary body—or whether they were true interstellar wanderers, caught by the passing Sun and bound into orbits around it, is still a matter for conjecture. Observationally, the unifying feature of nearly all comets is that they move in very eccentric orbits, swinging near the Sun—perhaps even closer than Mercury—at their perihelia, but reaching the realms of the outer planets when at aphelion. Only when they are near perihelion, shining both by reflected sunlight and by the luminosity of their excited gases, are they bright enough to be seen. It may well be that the solar system contains multitudes of cometary bodies that never approach the sun close enough to be seen at all, and that eccentric orbits are really the exception rather than the rule; but we can never hope to detect such ghostly bodies with our present observational means, unless they happen to pass very close to the Earth.

Observing comets

Although some comets have been observed moving in orbits that will take them forever beyond the Sun's pull, and so can never return, the great majority of observed objects move in closed orbits, and their returns to perihelion can be predicted. Most of these orbits are of such vast extent that a circuit takes thousands of years, but there is a group of comets with periods of less than a century. Their orbits are well known, and predictions are issued well ahead of their reappearance, so that they are picked up when still very faint. These regular comets are, in most cases, too faint to be of much interest to binocular observers even when near perihelion; Halley's comet, due back in 1986, is the brightest example. New discoveries, however, are usually made accidentally on photographic plates exposed for some other purpose, or by patient comet-hunters, and many of these objects are sufficiently bright to be of interest to the modestly equipped amateur, although they may appear as no more than nebulous blurs, without any trace of the striking tail exhibited by the rare brilliant comets. However, the amateur can do very useful work in following any comet within his range for as long as possible, estimating its magnitude and watching for any features, such as jets or clouds of matter, that may issue from the main body or "coma" of the comet. Some comets are quiescent and show little variation in brightness that could not be predicted on the basis of their changing distance from the Sun and Earth, but others have exhibited remarkable activity, flaring by a factor of thousands of times in a matter of a couple of days!

The first thing for the would-be comet observer to do is to subscribe to an information service that sends out details of new discoveries, with preliminary predictions or "ephemerides" which give their track across the sky. The main source of information is the card service of the Central Bureau for Astronomical Telegrams (Smithsonian Astrophysical Observatory, Cambridge, Massachusetts), but many national societies send out their own circulars. The comet's course can then be plotted on a star chart, and it will at once be obvious if it is well placed for observation. Most new discoveries are made before perihelion, possibly not long before the comet moves too near the Sun to be observed, and a planisphere will indicate whether the comet is too low in the sky, or too near the twilight glow, to be worth seeking.

If the comet is of the 7th magnitude or brighter, useful brightness estimates can be made with a pair of 8 × 30 binoculars if the sky is dark; but moonlight, or twilight, will seriously affect the visibility of a fugitive object, and even a large instrument may, in a bright sky, be unable to show an object that is easily seen with a small aperture under good conditions. In making estimates of a comet's brightness we are usually seeking to find its *integrated* magnitude, which corresponds to the brightness it would have if its nebulous disk were condensed into a starlike point of light. To do this, nearby stars are put out of focus so as to match the comet in size, and its brightness relative to its defocused neighbors is estimated as in normal variable-star practice (page 79).

If the comet is bright, the comparison stars can be identified from *Norton's Atlas* and their magnitudes looked up in an appropriate catalogue (see Appendix 2). In the case of comets near or beyond the naked-eye limit, a field sketch, identifying the comparison stars, should be sent with the observation to the appropriate authority.

It will be seen from this that the observation of comets by amateurs, using very modest equipment, can be of great use. Indeed, the official announcement cards referred to above frequently carry notes on amateur observations of the brighter comets, since there are too few professional observatories around the world to maintain a constant watch on these objects.

Observing techniques

A few words on the procedure to be adopted when observing faint objects, such as most comets (unfortunately!) are, will be in order. The human eye is an extraordinary device, developed over millions of years to be able to serve its owner in conditions ranging from noonday brilliance to the darkness of a moonless night—a brightness factor of perhaps a million times. Clearly, such a range of intensity cannot be accommodated merely by varying the size of the iris, which expands from a diameter of about 2mm in daylight to perhaps 8mm at night: a range of brightness transmission of only 16 times. Instead, under conditions of low light intensity the sensitive retinal surface, on which the image of the object being viewed is formed, becomes bathed in a substance known as "visual purple." This visual purple has the effect of increasing the sensitivity of the nerve endings far beyond their normal daylight state.

Visual purple does not act instantaneously, and a person going from a brightly lit room into a dark garden is, initially, practically blind. Gradually the retina increases in sensitivity, and images come into view where all was previously black. The retinal state of an observer seeking faint astronomical objects is a critical factor in the success of his observation. At least ten minutes should elapse before any delicate work is attempted, and this can profitably be spent in looking round the sky, seeing what constellations are visible, and searching for a possible bright nova (see page 138). Only after this time should serious attempts at critical observation be made; and, should the object still not be seen, it should not be dismissed as "invisible" until some minutes more have been spent in searching the area. If street lights or lighted windows are troublesome, hiding the head under a blanket or cloth, with the instrument poking through, will work wonders. It has been proved that dark adaptation continues to improve for several hours, which is why every attempt should be made to avoid having to go back into the house to search for some forgotten item once the process has started.

Sometimes, the observer will suspect a faint star or comet but be unable to hold it definitely enough to confirm its presence. If this happens, wave the binoculars slightly so that the field objects move around; it frequently happens that the motion of a faint object on the retina makes it clearly visible, although it is barely detectable, or even invisible, when stationary. The technique of "averted vision," or looking at one part of the field while directing the attention to the point under examination, is well known.

Sweeping for comets

At first sight it might appear that the chances of an observer discovering a comet with the usual hand binoculars are very small, inasmuch as an object bright enough to be seen in such an instrument could hardly have escaped the searches of well-equipped amateur and professional observers. However, this is not always the case. The bright Comet Mrkos of 1957 was discovered moving out from the vicinity of the Sun as a naked-eye morning object. Two comets discovered in 1975, well away from the sun and visible in a dark sky, were picked up by the writer, using 12 × 40 binoculars, on the night following their announcement. Therefore, it always pays to check the atlas whenever a nebulous object not positively identifiable is sighted.

Comet nomenclature

When a comet is discovered, it is identified by the name of its discoverer (or names, if found independently) together with the year and a small letter indicating its order of detection in that year. Thus we have Comet Kohoutek, the sixth comet found in 1973 (this includes recoveries of known objects), known as 1973f. Later, the small letter is replaced by a permanent Roman numeral designation, indicating the year and order of perihelion passage. Comet Kohoutek was the first comet to pass perihelion in 1974, and is now known as 1974I.

METEORS

The connection between comets and meteors was demonstrated, strikingly, by the celebrated comet discovered by Biela in 1826. At the 1845 return it was observed actually to split into two distinct units—an unprecedented event. In 1852 (its period was about six years) the two components were still within range of each other, but at the next two returns nothing at all was seen. It should have returned again in 1872; instead, there was a magnificent meteor shower at the time when the comet should have passed close to the Earth. The inference is obvious: comets consist mainly of meteoric particles, and it is now known that many of the principal meteor showers are associated with the orbits of known comets.

A meteor, in its own right, is a tiny granule of matter no bigger than a grain of sand, circling the Sun in a normal orbit. If left to its own devices it would keep on eternally, but there is just the remote chance that it will find itself on an unavoidable collision course with the Earth. When this happens—and millions of meteors are captured daily—it shoots through the atmosphere at such a velocity that the air becomes luminously hot. This is the "meteor" that we see with the eye.

Meteor showers

On almost any night of the year, provided the sky is really dark (and towns are no place for comet or meteor work), six or eight faint meteors will streak at random among the stars. These are called "sporadic" meteors. But at certain times of the year, notably August and December, definite showers occur. Perhaps "shower" is an overambitious word; it simply means that the meteors all appear to come from the same part of the sky, and they occur when the Earth passes through a swarm of meteoric particles.

This is where their relationship with comets comes in. As a comet pursues its path round the Sun a certain amount of its meteoric content gets left behind, so that there is a trail of debris circling perpetually in its orbit. If the Earth's orbit chances to cut the comet's path it will experience a sudden increase in meteoric activity. This produces a shower, and since the two orbits are fixed in space it must obviously occur at the same time every year.

Meteor shower densities can vary from about two to sixty naked-eye meteors per hour. If the trails belonging to a particular swarm are plotted on a star map and traced back, they will be found to radiate from a certain region of the sky: the "radiant." This is actually an effect of perspective; the meteors are traveling in parallel paths, but they are so far apart that their trails appear to widen as they approach.

The constellation in which the radiant is situated gives the shower its name. The August meteors radiate from Perseus and are called the Perseids; Gemini provides the December shower, which are known as the Geminids. Sometimes the same constellation plays host to two or more showers at different times of the year, in which case the brightest star near the radiant provides the distinction. For example a shower on January 17 radiates from κ Cygni; the meteors are called the κ Cygnids to distinguish them from a shower which occurs round about the end of July. The radiant star in this case is α, and they are called the α Cygnids.

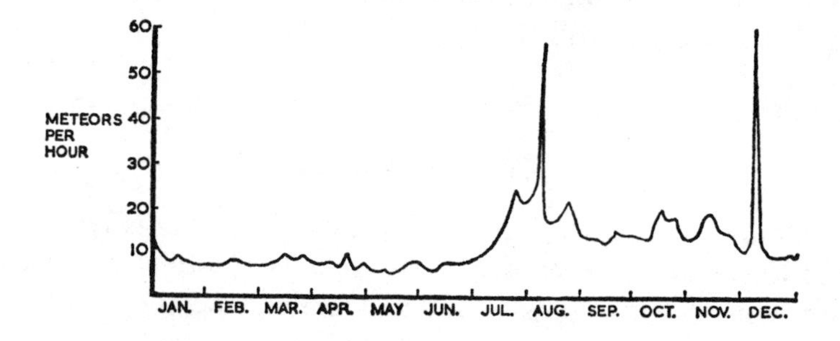

Fig. 19. *Meteor rates throughout the year*

The figures refer to real, not zenithal, hourly rates, as seen from England. Though actual results depend enormously on observing conditions (it is therefore perhaps misleading to quote specific rates) the graph does at least show that autumn is the best time of the year for casual meteor observation. The two main showers are the Perseids and the Geminids.

Observing meteor showers

Twenty years ago meteor observation was almost entirely in amateur hands, and the universal instrument was the naked eye. What observers did was to lie on their backs gazing up at the stars, noting the points of appearance and disappearance of each meteor and quickly plotting it on a chart. By this means a great deal of information could be obtained about the nature of the radiant (whether it was sharply defined or spread over a large area) and how it moved during the duration of the shower. Since the Earth is moving through the swarm all the time, it is to be expected that the meteors will appear to change direction slightly.

But times have changed, and professional astronomers have now entered the field armed with photography. This has become possible only very recently, with the advent of super-fast emulsions and wide-aperture cameras, but the advantages are obvious: the trails can be recorded far more accurately than by the best visual plotting, so that the derived results are more accurate. It does, however, suffer from one drawback. Photography can record only the brighter meteors, so that knowledge of the fainter members is still more or less in amateur hands.

With this in mind, the naked-eye observer has a simpler but also less spectacular job. All he has to do is count the meteors as they appear, noting the change of number that are recorded during fixed intervals throughout the shower. These intensities, when plotted on a graph, give an indication of the variation in strength of the shower and the epoch of maximum activity.

The Perseids, one of the most dependable showers, can be taken as an example. Maximum activity usually occurs early in the morning of August 12, so to cover the main period observations should, ideally, be continued from dusk on August 11 to dawn (like comet-seeking, meteor observation needs patience). Outliers of the Perseid swarm are first seen at the end of July and continue for a fortnight or so after maximum, but it is only the phase of greatest intensity that is spectacular.

When the sky darkens, Perseus will be low in the northeast, and attention is therefore concentrated on this part of the sky. A record is then kept of all Perseids seen in each complete hour: 21^h to 22^h, 22^h to 23^h, and so on. It is usual, when submitting any astronomical observations, to reduce the local Standard Time to U.T. (Universal Time), but this can be done later. During the first hour, perhaps only ten meteors will be seen, but the intensity will increase as the night wears on.

When observation is over it is a good idea to plot a graph of hourly rate against time (Fig. 20). The result will perhaps be a steadily rising curve which drops down again late in the night, and from the graph in question it appears that maximum activity occurred at 1^h U.T. on August 12. These results are of course fictitious, and the curve is much neater than any actual curve is ever likely to be, but at least it illustrates the method.

Correcting the rate

In placing the time of maximum, however, an important factor has been forgotten. Perseus has been rising steadily throughout the night; it is still rising at dawn. This means, of course, that the altitude of the active area of the sky has also been increasing. At 21^h the radiant is only 27° above the horizon; at 23^h it is 37°; at 1^h it is 51°; and at 3^h, just before dawn, it is 66°. The inference is obvious; the increased altitude is going to mean firstly that a fair percentage of the early meteors occur below the horizon, and secondly that they have to be seen through a thicker layer of air, so that the fainter ones are drowned altogether. We know that

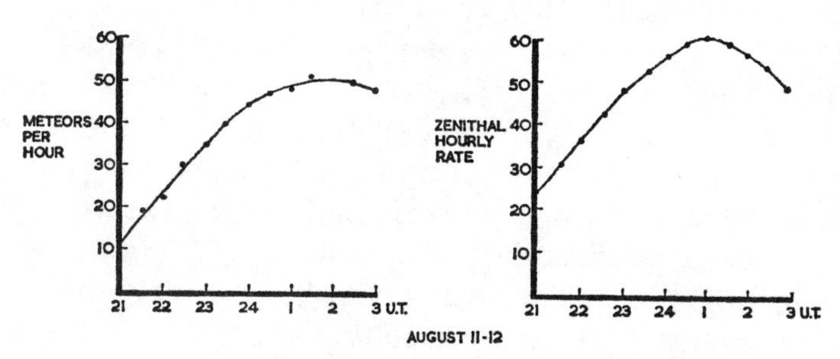

Fig. 20. *The Perseid meteors: apparent and real rates*

In practice, the Z.H.R. curve is never as neat as the one shown here!

stars appear dimmer near the horizon, and the same naturally applies to meteors. Hourly rate is therefore intimately connected with the radiant's altitude, and there must be some means of correction for this. These means are embodied in the following table.

These factors give the amount by which the observed rate for the appropriate altitude must be multiplied in order to obtain what is known as the Zenithal Hourly Rate. This is the rate always quoted in meteor shower predictions: it is the number of meteors that would be observed per hour were the radiant exactly in the zenith.

The hourly-rate graph can now be corrected. For example, we see that the rate at 27° is only half of the Z.H.R.; it is therefore doubled. When all the points are treated in this way the resultant curve looks (or should look!) like that in Fig. 20. Not only is the observed curve fictitious, but it places the time of maximum activity nearly an hour too late, the result of the radiant rising ever more favorably into the sky.

Meteor observation is one field of astronomical research in which cooperation of observers is almost essential if an amount of data suitable for analysis is to be amassed. The American Meteor Society collates observations by amateurs, and most national societies include a section devoted to this popular branch of observing.

Table 8 *Correction for Zenithal Hourly Rate*

This table gives the factor by which an observed hourly rate, with the radiant at a given altitude, must be multiplied to obtain the Z.H.R. It is a theoretical table only, applying to excellent sky conditions; moonlight or slight haze may easily double or treble the factor if the radiant is low, so it must not be treated too dogmatically.

ALTITUDE	FACTOR	ALTITUDE	FACTOR
2°6	10·0	34·°5	1·7
8·6	5·0	42·5	1·4
14·5	3·3	52·2	1·25
20·7	2·5	65·8	1·1
27·4	2·0	90·0	1·0

Corrections for intermediate altitudes can be obtained by interpolation.

There are many recognized meteor showers, and they all have interesting peculiarities. A list of the main ones, together with details, is given in Table 9.

Apart from the weather, which rarely cooperates on these nights, meteor observation is severely handicapped by strong moonlight. A brilliant Full Moon will drown all but the brightest stars, and it is naturally futile to search for a faint shower under such conditions. Moonlit periods vary each year, affecting some showers and leaving others confined to the favorable fortnight, so advantage should be taken of this. For instance, the Perseid maximum in 1957 had to cope with a Full Moon, whereas in 1961 the Moon was near the Sun and had no effect at all.

Not all maxima occur during the dark hours, of course, and not all are as sharp as the Perseids'. The whole extent of the Quadrantid shower (January 3–4) is confined to a few hours, even though the Z.H.R. is sometimes over 100; whereas the Taurids are about equally intense all the time with a rate of 10. Leap year adjustments cause the precise time of maxima to vary each year.

Fireballs

While awaiting the smaller fry of meteor showers, occasional really brilliant objects known as "fireballs" will be seen. There is no strict definition of a fireball, and people frequently have rather strange ideas about them, confusing them with thunderbolts and other terrestrial phenomena. In actual fact a fireball is simply a large meteor—perhaps as big as a small pebble—and most of them are sporadic.

When a fireball bursts into view, note the beginning and end of the trail as accurately as possible; to do this requires a good knowledge of the sky, but this will come with frequent meteor-watching. If it can then be identified with sightings by other observers, its height and actual flight path can be calculated by a simple piece of triangulation.

The principle is simple enough. Suppose that observers X and Y note the beginning point of a fireball's track at a certain altitude and azimuth, and observe its end at another altitude and azimuth (Fig. 21). The meteor's height and position above the ground can

Table 9 *Some Important Meteor Showers*

There are many discernible showers, but only a small proportion give anything in the nature of a notable display, and only those likely to give maximum rates of the order of 5–10 or more meteors per hour to a well-situated observer are listed here. The figure in parentheses refers to the likely maximum Z.H.R. Some showers come to peak within the space of a few hours, and, since leap-year adjustments affect these timings, a current astronomical almanac should be consulted.

DATE	SHOWER	REMARKS
Jan 1–5	Quadrantids	A very sharp maximum around Jan 3, sometimes occurring in daylight. The shower derives its name from the old constellation of Quadrans Muralis; the naked-eye star nearest the radiant is θ Boötis. (100?)
Apr 19–24	Lyrids (April)	Very swift meteors, max Apr 21. (12)
May 1–8	η Aquarids	Very swift. Best seen just before dawn, max May 5. (20)
Jun 10–21	Lyrids (June)	Max Jun 16. (8)
Jun 17–26	Ophiuchids	Max Jun 20. (8)
Jul 15–Aug 15	δ Aquarids	This strong shower overlaps the early Perseids, giving much meteor activity in early August. Max Jul 28. (35)
Jul 15–Aug 20	Piscis Australids	Max Jul 31. (10)
Jul 15–Aug 25	α Capricornids	Slow meteors & fireballs. Max Aug 2. (10)
Jul 15–Aug 25	ι Aquarids	Max Aug 6. (8)
Jul 25–Aug 18	Perseids	Swift, bright meteors; a very prominent northern shower. Max Aug 12. (65)
Oct 16–26	Orionids	Swift meteors. Max Oct 21. (35)
Oct 20–Nov 30	Taurids	A long-lasting shower of slow meteors, with two radi-

		ants separated by 8° in dec. Max Nov 8. (15)
Dec 7–15	Geminids	Medium speed, fireballs; many telescopic meteors. Max Dec 14. (60)
Dec 17–24	Ursids	Max Dec 22. (10)

then be worked out once the distance and direction between the observers' sites is known. Given its direction of flight through the atmosphere, the path through space that the fireball was following, and hence its original orbit round the Sun, can be calculated if the duration of the meteor is known. It is here that the greatest errors tend to appear, and the observer should, as part of his basic training, learn to count seconds in his head as accurately as possible, perhaps in the manner "ONE thousand, TWO thousand, THREE thousand," and so on, and he needs to have the presence of mind to start his mental clock counting the instant a fireball bursts into view, as well as noting the position among the stars where it appears and vanishes. Even an untimed report, however, can be used for obtaining a provisional orbit, since all meteor velocities through space approximate to each other at the Earth's distance from the sun.

A really good fireball observation is of a value commensurate with its rarity. It often happens that the most interesting objects are seen by the least competent observers, and vague references to direction and brightness are of little use beyond establishing that the phenomenon occurred. The temptation to think that the object he has just seen was so bright that many reports must have been sent to the national center, and it is not worth adding to the pile, is natural to the novice, but is sadly mistaken. His report may be just the one needed to add certainty to a doubtful path. If the observer feels that his observation is a good one, and can pinpoint positions on the track to within two or three degrees in the sky, it should certainly be submitted. Reduction to altitude and azimuth will be done by the computer: what is wanted is the position in the sky relative to the stars. Just one point—beginning, end, or somewhere along the track where it passed very close to

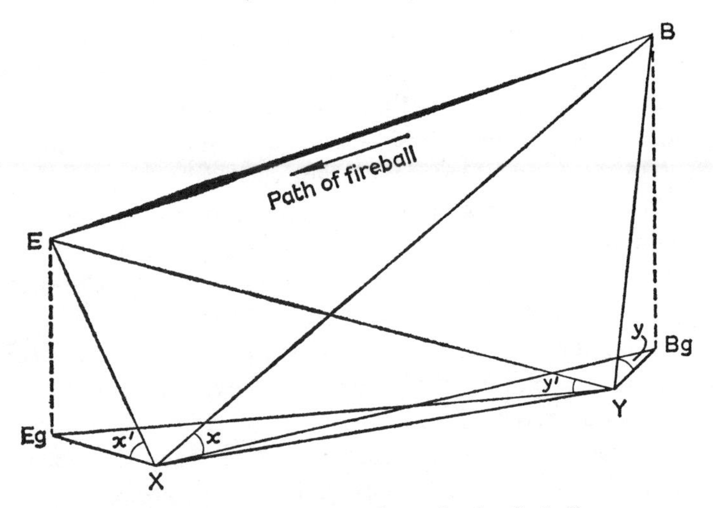

Fig. 21. *Determining the path of a fireball*

If the positions of observers X and Y are known, the points Bg and Eg directly beneath the beginning and end of the track can be obtained by combining the observed azimuths of XB and YB, and XE and YE. The beginning and end heights can be derived by combining the observed altitudes x and y, and x′ and y′.

an identifiable star, or between two stars—will be of use, but two or three points will be invaluable. Should the fireball be large enough to drop a meteorite, a good observation may help to define the path accurately enough for a possible landing site to be identified and searched.

Most fireballs leave a faint trail of ionized gases, known as a *train,* in their wake, and it is here that binoculars prove useful, because the behavior of this train—its drifting or breaking-up—is affected by the little-known high-altitude winds. Sometimes trains last for several minutes, and an account of what happens will be of value.

AURORAE

Also to be included in the naked-eye category are aurorae, the strange glows that sometimes light up the northern sky in displays ranging from greenish hazes to vivid red curtains. Aurorae are caused by very short-wave particles emitted from sunspots exciting some of the gases in the upper atmosphere, and they are therefore strongly dependent on the sunspot cycle, which, at present, is in a period of maximum activity.

Aurorae center around the north magnetic, not geographic, pole, which at the moment is some 10° west of true north. When a display is seen, first of all note its bearing with reference to a compass or the pole star; then note the time at which anything unusual happens, such as a sudden brightening or the appearance of any unexpected activity. A general description of the display should also be given.

Just occasionally really great displays occur, especially in higher latitudes which are nearer the main auroral region, and these can take amazing forms. Sometimes the light is rayed in a huge arc, shimmering spectrally and casting beams well into the southern sky. Other aurorae take the form of a huge shimmering curtain, apparently waving in some cosmic breeze.

Binoculars are useless for observing aurorae, for their magnification is a positive disadvantage; it blurs the outlines and dilutes the light. They are, however, useful for picking up stars to mark the limit of the display, and notes on its extent at different times are helpful.

For measuring the size of the auroral glow some sort of angular scale must be used. A useful rule to remember is that a foot ruler held at arm's length subtends an angle of about 24°—or 1° per half-inch. If a ruler is not available, from thumb to little finger of the outstretched hand is about 20°. This does at least give some sort of key, though if more accurate results are needed a measuring instrument like a quadrant could easily be made. This would also have obvious applications when measuring fireball trains.

THE ZODIACAL LIGHT

Really vivid aurorae can sometimes be seen from built-up areas, but there is another glow in the sky—a permanent one—which is faint and very difficult to see. This is the Zodiacal Light, caused by sunlight reflected from meteoric particles far out in space. As might be expected it keeps to the region of the ecliptic, appearing as a narrow cone tapering to an apex about 60° away from the Sun.

Because it is so faint it is best seen when the ecliptic makes its greatest angle with the horizon. The law is naturally the same as for Mercury and Venus: a spring evening after sunset, or an autumn morning. The sky, needless to say, must be absolutely dark, and the clearest conditions of air are essential for a good view.

Once again the tropics have the best of it, for the cones, both evening and morning, rise almost vertically into the sky, making the Light a normal sight. It is rather elusive from the northern States, where it will probably be seen most easily on a September morning, in the last minutes of darkness before dawn starts to break.

Although the Zodiacal Light is best seen from low-latitude sites, at which the ecliptic makes the maximum angle with the horizon, it is a far from difficult object in temperate zones when the sky is clear. Much harder to see is the Zodiacal Band, a very faint extension that circles the sky, joining the apices of the two cones. At the midpoint of the band, opposite the sun in the sky, is the glow known as the *Gegenschein* which, although brighter than the band and 20–30° across, is still very difficult to see. It is best to look for it when it is projected upon a part of the sky devoid of bright stars, as occurs in the northern autumn when it lies in the Pisces-Cetus region.

☆ 6 ☆

Beyond the Solar System

Pluto: $5\frac{1}{2}$ light-hours. The nearest star: $4\frac{1}{3}$ light-years. The farthest our optical telescopes can see into space: about 4,000 million light-years,* possibly more. This is the universe, a tiny local part of which we see every time we glance at the sky on a clear night.

This is a practical book, but practice is not possible without at least some understanding of the objects being observed; a star, for instance, is fascinating not through its own beauty but because of what we know about it, and the same is true of spectacular objects such as nebulae or star clusters. So a lightning survey of our astronomical surroundings will not be out of place.

The best way is to use a scale model. Shrinking the Sun to the size of a basketball (when the Earth becomes about 2mm across), we cast around for the nearest star. If we are searching on foot it will take rather a long time, for we shall have to walk no less than four thousand miles. Pluto we shall pass after only a thousand yards. Clearly this is an impracticably large scale if we want to construct the Sun's immediate neighborhood, for the Earth itself could accommodate only a few stars.

*The light-year is an evocative unit, but astronomers always measure large distances in *parsecs* ("parallax of one second"), which refers to the distance at which the radius of the earth's orbit would subtend one second of arc. This unit is equivalent to 3·26 light-years or 19·17 million million miles (or 19·17 trillion miles).

So the scale must be shrunk again, letting the distance between the Sun and the nearest star be one foot. We can then construct a cross-section through our local cluster or "galaxy" by drawing a circle with a diameter of six miles and filling it with dots one foot apart. Each dot is a star like the Sun, though in practice it would be impossible to make them small enough to fit the scale.

The Galaxy

The Galaxy is in the form of a spiral about 30,000 parsecs across, slowly rotating—one revolution takes over 200 million years, so it is quite inappreciable. From the side it looks rather like two fried eggs placed bottoms together, with a conspicuous bulge of stars and gas near the nucleus. Altogether, the Galaxy contains at least 50,000 million stars (some estimates are double this number), and most of them are separated by the same order of distance that splits the Sun from its neighbor.

This, our local star system, is by no means the only one. Galaxies are scattered throughout space; they are units of the universe in the same way that sand grains are the units of a seashore. And the simile is appropriate enough, for the number of galaxies detectable with our largest instruments runs into the thousand million. Only one, however, is close enough to be at all conspicuous; it appears as a milky blur in the constellation of Andromeda. All the night-sky stars are simply stars in the Sun's neighborhood, belonging to our own Galaxy.

The Galaxy contains interesting objects other than stars. These are the nebulae, vast clouds of gas which will one day give birth to stars. One is visible with the naked eye; it is in Orion, below the famous belt, and several more can be picked up with binoculars, though it must be admitted that they are not spectacular. We also find shells of luminous gas believed to have been emitted from dying stars; these are known as *planetary* nebulae because of their somewhat disklike appearance, but most are too faint to be observed with binoculars.

In addition to these bright nebulae there are colossal obscuring masses of dark gas and dust, which cannot be seen in themselves but are made visible by the stars they cut off from sight. Several examples can be seen in the Milky Way (which is our view through the thickest part of the Galaxy), appearing as black vacancies among the stars, and they show up much more prominently on long-exposure photographs. From our point of view the most important dark clouds are those which lie between us and the nucleus of the Galaxy, in the direction of Sagittarius, because the star density here is so thick that were it not for these tremendous filters their brightness would light up the sky far more effectively than the Full Moon.

THE STARS

In addition to the nebulae there are three types of stellar object that can come under scrutiny: double stars, variable stars, and star clusters.

Double stars

A double star is a star which appears single with the naked eye, but double, or possibly multiple, with a telescope. This distinction naturally depends on the kind of instrument employed, for a double which may tax the powers of binoculars will probably be too easy to be worthy of the name when an astronomical telescope is turned upon it.

The distance between the components is measured in terms of angle, the unit of which is the second of arc (″). A second is $\frac{1}{360}$ part of a degree, so it is obviously a very small angle; the width of a half dollar 4 miles away subtends 1″, and the Moon's diameter is normally about 2000″, or 33′ (minutes of arc, or $\frac{1}{60}$ of a degree).

It takes a fair-sized telescope to resolve a double star whose components are only 1″ apart, and a great deal depends on their brightness; if one star is much brighter than the other its glare will make separation more difficult. Normally, 8 × 30 binoculars cannot resolve a double much closer than 30″; the unaided eye can split stars down to about 4′ apart under very favorable conditions.

Physically there are two distinct types of double star: the optical double, and the binary. An optical double is simply two stars which happen to lie in almost exactly the same line of sight, so that they appear to be almost touching, while in reality they are in no way connected. A binary system, on the other hand, consists of two stars revolving around their common center of gravity. The periods are always long, and may stretch up to hundreds of years, in which case it may take several decades to notice their motion. Amazingly enough, fully a quarter of all the stars in the Galaxy are members of binary systems, and it is perhaps a pity that the Sun chances to be a lone wanderer in space—strictly from the aesthetic point of view!

Observing double stars

The observer will learn from experience the separations and brightnesses of the double stars that he can expect to divide with his equipment. The resolving power of a telescope can be applied with significance only to pairs of stars of roughly equal brightness, and neither too bright nor too faint; a mag. 5 pair would be about the optimum for the powers of an 8 × 30 binocular. If one component is very much fainter than the other, the resolving limit may be doubled or trebled because of the glare produced by the brighter star. In order to document a double star, so that the observer can form a mental picture of its appearance, the magnitudes of the components, their separation in seconds of arc, and the position angle (see page 15) of the fainter star with respect to the brighter one, have to be known. Thus, if we were to observe the star 56 Andromedae (see page 94), we would find the information written as 5·8, 6·1; 300°; 190″. This means that the magnitudes of the components are 5·8 and 6·1; that the stars are 190 seconds of arc apart (about $\frac{1}{10}$ of the Moon's apparent diameter), and that the position angle of the fainter star from the brighter is 300°, which means that it lies to the WNW.

Variable stars

However, we can think it fortunate that the Sun's output of light and heat seems to have been fairly stable, for there is a class of star whose light output is anything but regular, known as variable stars. Once again there are two types. In one, the main one, they are genuinely variable; drastic changes in the star's constitution, which may or may not be regular, cause it to brighten and fade periodically. In the other group the stars are nothing more than binary systems seen edge-on, so that the components mutually occult each other. These "eclipsing variables" usually have very short periods of just a few days, and they can usually be predicted with great accuracy, although the variations of some of these stars do steadily wander from the predictions.

The genuinely variable stars, known as "intrinsic variables," can be divided into a number of classes. There are pulsating cool stars, much larger than the sun, which, because of their lower temperature, are reddish in color. Many of these have periods of fluctuation of 300 days or more, and are known as Long Period Variables (LPVs). The change of brightness of these stars between maximum and minimum can be of the order of thousands of times; a number are visible with binoculars when near maximum, but most require fair-sized telescopes for observation at minimum.

Allied to the LPVs are the semiregular variables. These, too, are giant cool stars, but their periods of fluctuation do not follow a constant law, and their brightness ranges are less. A number of these can be observed with binoculars throughout their cycles, and since these stars have been less well studied, observations can be really useful.

Truly irregular stars also exist. Some of these, such as Betelgeuse (α Orionis), can be considered as an extension of the semiregular family, but others, such as ρ Cassiopeiae, are subject to bursts of activity at intervals of perhaps decades. Stars as bright as this can be followed well with the naked eye. Then there are the curious and rare stars called after their prototype, R Coronae Borealis, in the small but easily located constellation of the Northern Crown. These spend most of their life at maximum brightness (R is normally just visible with the naked eye), but at intervals of several years, and without warning, they drop precipitously; at its minimum of 1972 I could only just glimpse it with a 22cm telescope. The binocular observer is, therefore, well equipped to announce the onset of a minimum of this strange star. The other principal class of variable star, the so-called dwarf novae, which suffer fairly regular outbursts, are all too faint to be of interest to observers with small apertures.

Observing variable stars

Variable stars are estimated by comparing their brightness or "magnitude" (page 83) with that of nearby stars whose magnitudes are known. The easiest way is of course to find one of the same brightness, since the magnitude can then be derived without ambiguity. Unfortunately this convenient state of affairs rarely occurs.

In this case select two stars which are slightly brighter and dimmer than the variable, and estimate its position in the sequence. Suppose, for example, that the magnitudes of the comparison stars are 5·4 and 5·8. If the variable is midway in brightness, call it 5·6; if it is closer to one star than the other note it as 5·5 or 5·7 as the case may be.

When making repeated observations of a variable, try always to use the same comparison stars. This will tend to eliminate errors due to color differences and other effects. Obtaining satisfactory magnitudes for these stars may prove a problem, since there are many catalogues giving their own private values. It is therefore important to quote the source, if a standard chart, giving comparison star magnitudes, is not available. A further discussion of this work appears on page 131.

Star clusters

Star clusters, too, form themselves into two separate types: open and globular. Open clusters are typified by the Pleiades; the stars are well scattered, and binoculars can resolve them without much difficulty. Globular clusters, however, are far more compressed, and even the brightest example available to northern observers—the cluster in Hercules—appears as little more than a circular glow with a low magnification.

Catalogue of deep-sky objects

There have been many catalogues of nebulae and clusters. The first reasonably comprehensive one was drawn up by Messier, an Italian comet seeker, in 1781. Because it was early it contained all the bright objects, and his catalogue references, preceded by the letter M, are still used today. The galaxy in Andromeda, for instance, is M.31. Messier included 103 objects in his list, and they are nearly all visible with binoculars. The most comprehensive listing of nebulae and clusters is the *New General Catalogue* (1888), which contains over 7,000 objects that are identified by the letters N.G.C. A supplement to the N.G.C., the *Index Catalogue* (I.C., 1904 & 1908) contains few objects of interest to binocular users.

Numerous catalogues of double stars have been issued, but most binocular pairs are bright enough to have been included in the original naked-eye listings (see page 81).

THE CONSTELLATIONS

A constellation is simply a specified area of the sky, and most of the northern groups had their rather romantic origins well back in the dim mists of history. They were carefully fitted in with mythological traditions, and by comparison the more modern additions, dating from the seventeenth century, sometimes seem very uncelestial; the southern hemisphere, especially, is decorated with utilitarian articles such as a compass, a furnace, and a painter's easel. But astronomy, mercifully, is more conscious than any other science of its ancestry, and the sky today reflects this confusion of misguided imagination without any attempts at stultifying it with order and logic.

The year 1930 is an important date in constellation history. Before then boundaries had been rather nebulous, and some star

1. The Moon, 4 days old. The larger craters shown in this and the following photographs are easily detectable with binoculars, and can be identified by means of the outline map. *(Lick Observatory)*

2. The Moon, 7 days old. *(Lick Observatory)*

3. The Moon, 10 days old. Compare the position of the Mare Crisium with that shown in Plate 1. The two photographs were taken, respectively, near eastern and western libration. Note also the Sinus Iridum on the terminator. *(Lick Observatory)*

4. The Moon, 14 days old. The rays from the southern crater Tycho are now the most prominent feature. *(Science Museum, London)*

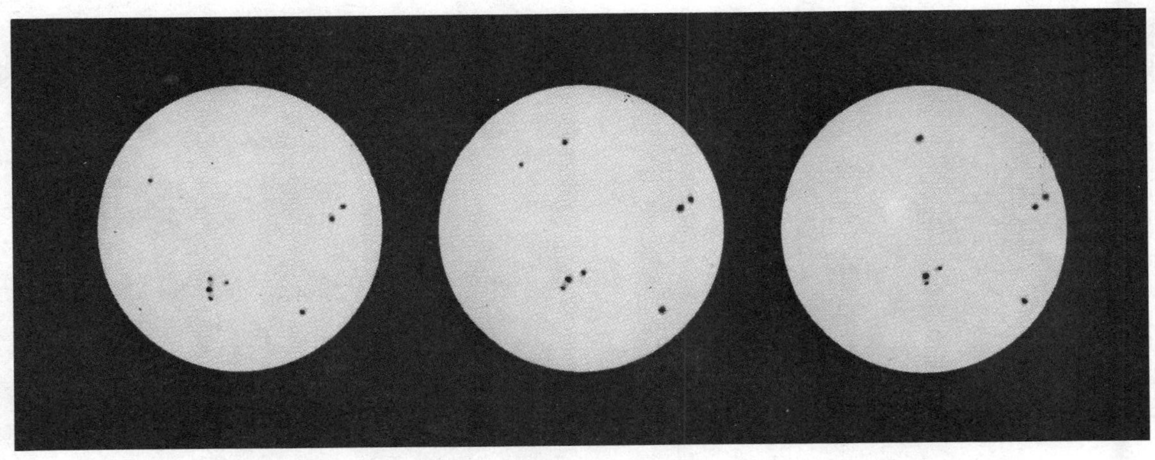

5. Sunspots. Three drawings made by the author on successive days in 1956. 8×30 binoculars were used, with a deep green filter in front of each objective. The drift of the spots, due to the rotation of the Sun, is very obvious.

6. Comet Austin, 1982g. Discovered by a New Zealand amateur, this comet was a conspicuous binocular object in the summer of 1982. (*photograph by R. W. Arbour*)

7. The Hyades. This open cluster, which is near the line of sight of α Tauri (Aldebaran), below and left of center, is a fine binocular group. *(photograph by the author)*

8. The Region of γ Cygni. An area of sky measuring about 5½° × 7½° around this bright star; the arrow indicates the highly luminous star γ Cygni (page 105). North is towards upper left. *(photograph by the author)*

9. The Pleiades. The visibility of the fainter members of this group forms a good test of naked-eye vision. *(photograph by the author)*

10. The Andromeda Galaxy, M.31. The bright central region, and the two satellite galaxies (arrowed), can be detected with small instruments. North is to the left. *(photograph by R. W. Arbour)*

catalogues had disagreed as to which star belonged to what constellation; Flamsteed, the first Astronomer Royal, had unintentionally duplicated some stars in his catalogue of 1725, and there were many other similar instances. In this redeeming year, therefore, the International Astronomical Union set down definite constellation boundaries, and also issued a list of standard three-letter abbreviations which are now in universal use (Table 10).

Star identification

So many catalogues have been issued that star identification is, frankly, in a state of confusion. Luckily the brighter stars are catered for by two systems, those of Bayer (1603) and Flamsteed, and binocular work is unlikely to require reference to any of the more advanced modern catalogues.

Bayer used the Greek alphabet, and his general idea was to label the stars in each constellation in order of brightness; in the constellation Lyra, for instance, α Lyrae is the brightest, with β and γ coming second and third respectively. This system, of course, limits itself to the 24 brightest stars, and in any case Bayer was not systematic; he labeled Ursa Major's seven main stars in simple order of position, and there are other instances also. Nevertheless, the alphabetical listing is so well established that it will never be changed now.

When reference is made to the fainter but still naked-eye stars, Flamsteed's numbers are used. In his catalogue he numbered every naked-eye star in each constellation not in order of brightness, but in order of position, from west to east. A small constellation might therefore muster only twenty or thirty stars, while a rambling one might reach a hundred or more. Flamsteed numbered every star, including the ones Bayer had catalogued, but in these cases the Bayer letter is always used. This double system may appear confusing at first, but both are logical, and for amateur work they are preferable to modern catalogues which ignore constellations altogether and simply work round from west to east.

Bayer and Flamsteed were both northern observers, and the

Table 10 *The Constellations and Their Abbreviations*

Most of the abbreviations are simply the first three letters of the constellation name, but there are exceptions where this might cause confusion or duplication; e.g. Hya, Hydra; Hyi, Hydrus. The abbreviations also act as the genitive, e.g. α Hya.

Abbr.	Constellation	Abbr.	Constellation
And	Andromeda	Gem	Gemini
Ant	Antlia	Gru	Grus
Aps	Apus	Her	Hercules
Aql	Aquila	Hor	Horologium
Aqr	Aquarius	Hya	Hydra
Ara	Ara	Hyi	Hydrus
Ari	Aries	Ind	Indus
Aur	Auriga	Lac	Lacerta
Boo	Boötes	Leo	Leo
Cae	Caelum	Lep	Lepus
Cam	Camelopardus	Lib	Libra
Cap	Capricornus	LMi	Leo Minor
Car	Carina	Lup	Lupus
Cas	Cassiopeia	Lyn	Lynx
Cen	Centaurus	Lyr	Lyra
Cep	Cepheus	Men	Mensa
Cet	Cetus	Mic	Microscopium
Cha	Chamaeleon	Mon	Monoceros
Cir	Circinus	Mus	Musca
CMa	Canis Major	Nor	Norma
CMi	Canis Minor	Oct	Octans
Cnc	Cancer	Oph	Ophiuchus
Col	Columba	Ori	Orion
Com	Coma Berenices	Pav	Pavo
CrA	Corona Australis	Peg	Pegasus
CrB	Corona Borealis	Per	Perseus
Crt	Crater	Phe	Phœnix
Cru	Crux	Pic	Pictor
Crv	Corvus	PsA	Piscis Austrinus
CVn	Canes Venatici	Psc	Pisces
Cyg	Cygnus	Pup	Puppis
Del	Delphinus	Pyx	Pyxis
Dor	Dorado	Ret	Reticulum
Dra	Draco	Scl	Sculptor

Equ	Equuleus	Sco	Scorpius
Eri	Eridanus	Sct	Scutum
For	Fornax	Ser	Serpens
Sex	Sextans	Tuc	Tucana
Sge	Sagitta	UMa	Ursa Major
Sgr	Sagittarius	UMi	Ursa Minor
Tau	Taurus	Vel	Vela
Tel	Telescopium	Vir	Virgo
TrA	Triangulum Australe	Vol	Volans
Tri	Triangulum	Vul	Vulpecula

southern skies were not catalogued until more recently; but Greek-letter designations are, in most cases, applied to the brighter stars in each constellation.

Some of the brightest stars have received proper names: for example, the brilliant α Lyrae is usually referred to as Vega. The more common of these names are given in Chapter 7.

Variable stars have a special system of nomenclature. The brighter ones, whose variability might have been undetected when Bayer and Flamsteed drew up their catalogues, mostly retain their original references; thus α Orionis is a variable, and so is 15 Monocerotis. Other fainter variables have been given letters, according to the constellation, beginning with R and working through to Z, followed by a strange series of descending pairs of letters (RR-RZ, SS-SZ, TT-TZ, down to ZZ). We then ascend to AA-AZ, BB-BZ, and so on down to QQ-QZ (omitting J). If the number of variable stars in the constellation exceeds the 334 allowed by these letter combinations, we proceed with V335, etc.

STAR MAGNITUDES

Star brightnesses are classified in grades or "magnitudes," a misleading unit since it has nothing at all to do with their size. The magnitude scale had its rough beginnings with the Greek astronomer Ptolemy; he called the brightest stars "first magnitude" stars, the faintest, sixth magnitude. The invention of the telescope naturally brought fainter stars into view, the limit of the 200-inch being near the 23rd magnitude, while the whole scale has been tightened into mathematical lines. A difference of five magnitudes between two stars means that one is exactly 100 times as bright as the other, while one magnitude means a ratio of about 2½, the exact factor being $\sqrt[5]{100}$.

The brightest star in the sky is Sirius (α Canis Majoris), with a magnitude of –1·45. Two other stars are also sufficiently bright to carry negative magnitude numbers: Canopus (α Carinae) at –0·86, and Arcturus (α Boötis) at –0·06. Vega (α Lyrae), which shines near the zenith on autumn evenings in north temperate latitudes, has a magnitude of exactly zero. The constellation notes give the magnitudes of various stars in each group. Planetary magnitudes vary according to their distance from the Sun and Earth as they revolve in their orbits; Venus, always the brightest, can reach –4·4, and Mars can attain –2·8 at a perihelic opposition. The other planets are fainter. Uranus, lingering near the threshold of naked-eye visibility, is about 5·5, while Neptune is of the 8th magnitude.

In a first-class sky, 30mm binoculars should be able to reach about mag. 9, but it depends so much on visual acuity and other factors that there is no point in defining a limit.

CELESTIAL POSITIONS

When we wish to define the position of a celestial object, the method used is basically the same as that used by a geographer in defining a point on the Earth's surface: by latitude and longitude. The celestial equivalent of latitude is declination (Dec.); that of longitude is right ascension (R.A.). Dec. is measured in degrees, north and south, but R.A. is measured in hours, from 0 to 24, for a reason which will become clear.

The celestial sphere

The easiest way of understanding celestial positions is by inventing a couple of fictions (Fig. 22). First of all, suppose that instead of being scattered throughout space the stars are all equidistant, attached to a sphere whose center is the Earth: the "celestial sphere." Then, instead of insisting that the Earth spins from west to east, pretend that the celestial globe spins from east to west. There is a precedent for this; all motion is relative, and in absolute terms we can speak of the Sun and stars moving round the Earth without being accused of Ptolemaic tendencies.

So with the celestial sphere revolving round us once in 24 hours, it is easy to fix certain points on it. First of all extend the Earth's axis to cut it, which produces the north and south celestial poles. It so happens that there is a bright star very near the north pole, though the southern hemisphere is not so fortunate.

Similarly, by extending the plane of the equator, we can produce a great circle on the sphere: the celestial equator. A moment's thought will show that any star on the celestial equator must pass directly overhead to an observer on the terrestrial equator, while observers in other latitudes will see the equator inclined lower and lower in the sky, until at the poles it runs round the horizon and the appropriate celestial pole is overhead.

These references can now be used to define a star's position. Its latitude, or declination, is the number of degrees it is north or south of the celestial equator (north positive, south negative). R.A., however, is a little more complex. Lines of longitude can be

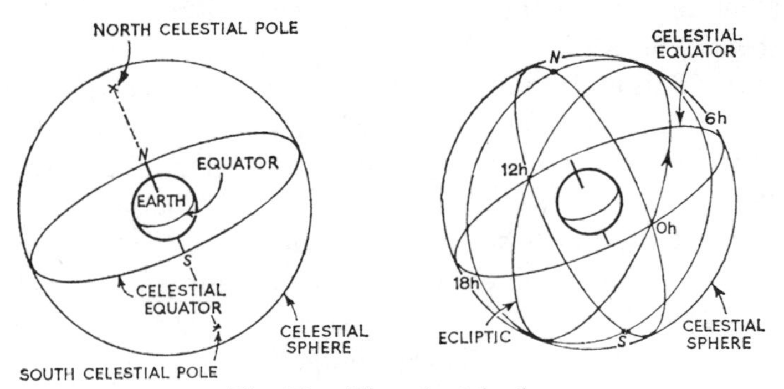

Fig. 22. *The celestial sphere*

drawn from pole to pole, but a starting point is needed on the celestial equator from which reckoning can start. Terrestrial longitude suffered from the same trouble until Greenwich was internationally recognized as 0°.

The Sun comes in here. It was shown on page 34 how it crosses the celestial equator at the two equinoxes; at the vernal going north, and at the autumnal going south. The spot at which it crosses going north, the vernal equinox, defines the zero of R.A., the starting point from which celestial longitude is measured. Going eastwards, the equator is divided into 24 equal sections of one hour each, since it takes 24 hours for the sphere to revolve once.

Figure 22 also shows the celestial sphere with the ecliptic drawn in, together with some lines of R.A. If we follow the annual path of the Sun, we find that when it is in 0° it is spring; in 6^h it is midsummer; in 12^h it is autumn, and in 18^h it is midwinter. The ecliptic is inclined to the equator at the Earth's axial tilt of $23\frac{1}{2}°$, so the four solar positions may be defined fully as R.A. 0^h, Dec. 0°; R.A. 6^h, Dec. $+23\frac{1}{2}°$; R.A. 12^h, Dec. 0°; and R.A. 18^h, Dec. $-23\frac{1}{2}°$. Star positions are given in the same way, though of course they stay the same from year to year.

THE SEASONS

There remains the explanation of why different seasons provide different constellations. The answer lies in the simple reason that the stars can be seen only at night; as we move round the Sun, we see them from a slightly different angle.

Figure 23 should make it clear. Suppose a star is being observed. When the Earth is in position *A* the Sun is in the region of the star, and it cannot be seen. At *B* (three months later), the star makes an angle of 90° with the Sun and can be seen in the early morning. By the time the Earth has reached *C* it is more or less between the Sun and the star, which is due south at midnight, like a planet at opposition. After that it slips further into the evening sky (*D*) until the Sun moves into its region again.

From this it is obvious that we see the star in the same

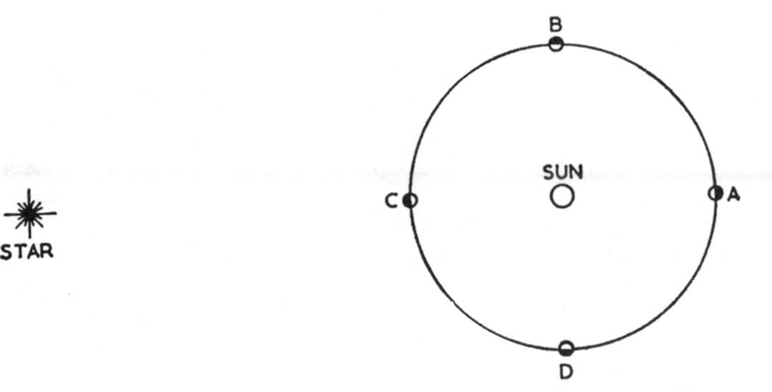

Fig. 23. *Seasonal shift of the constellations* (not to scale)

place a little earlier each night (since night and day are defined by reference to the Sun, *not* to the stars). The exact amount is easy to calculate: it is 24 hours multiplied by $\frac{1}{365}$, the sector of its orbit through which the Earth moves in one day. This comes out to about four minutes, which means that a star or constellation arrives at its previous position four minutes earlier each night.

SIDEREAL TIME

To complete this perhaps difficult section there is the matter of sidereal time or star time. The Sun, our civil timekeeper, returns to its old position in exactly 24 hours (neglecting slight variations). A star takes four minutes less, so that a sidereal day is in fact only 23 hours 56 minutes long, and the sidereal time at any instant is defined as the R.A. which is due south at that instant.

An example may help here. On December 21 the Sun is in R.A. 18^h, and at noon it is due south. So the sidereal time at noon on that day is 18 hours. By March 21 the noon sidereal time is 0 hours, the difference slipping away all the time. Table 11 gives conversions for 24^h (midnight) to sidereal time throughout the year.

A word or two needs to be said about what exactly is meant by "midnight." In practice, each point of longitude on the Earth's surface has its own instant of midnight, which occurs 4

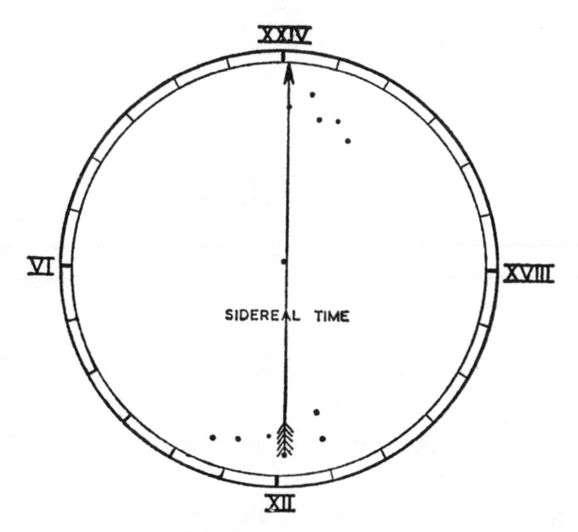

Fig. 24. *The cosmic clock*

Note that the celestial dial is numbered in an anti-clockwise direction.

minutes later for each single degree of westward longitude. For example, local midnight in New York City (about 73° W longitude) occurs some 16 minutes earlier than in Washington, D.C. (about 77° W longitude), and three hours earlier than in Los Angeles (about 118° W longitude). Since one time system cannot satisfactorily cover so large a spread in local time, the country has been divided into four single-hour time zones, centered along standard-time longitudes 75°, 90°, 105° and 120° west of Greenwich. The first zone, Eastern Standard Time (EST), is 5 hours slow on Greenwich Mean Time (GMT) or Universal Time (UT); Central Standard Time (CST) is 6 hours slow; Mountain Standard Time (MST) is 7 hours slow; and Pacific Standard Time (PST) is 8 hours slow.

It follows from this that "midnight" in any particular zone may differ by up to 30 minutes either way from the local midnight required for use in the conversion table. The difference between local midnight and standard midnight is set out below. By applying a 12-hour difference, the table can also give the time of local noon.

Table 11 *Sidereal Time at Midnight*

The table gives the sidereal time at midnight (24^h) on each of the dates mentioned. To obtain the S.T. for one hour earlier, subtract an hour; later, add an hour.

Sidereal time gains on G.M.T. at the rate of 3·9 minutes per day, from which it follows that S.T. at midnight on any given day is the same as that at 23^h a fortnight later; this, of course, is why the constellations are slowly displaced from east to west.

DATE	SIDEREAL TIME	DATE	SIDEREAL TIME
January 5	7^h	July 7	19^h
13	7½	15	19½
21	8	22	20
28	8½	30	20½
February 5	9	August 6	21
13	9½	14	21½
20	10	22	22
28	10½	29	22½
March 7	11	September 6	23
15	11½	13	23½
22	12	21	0
29	12½	29	½
April 6	13	October 6	1
14	13½	14	1½
22	14	21	2
29	14½	29	2½
May 7	15	November 6	3
15	15½	13	3½
22	16	21	4
30	16½	29	4½
June 6	17	December 6	5
14	17½	14	5½
22	18	21	6
29	18½	29	6½

OBSERVER'S LONGITUDE (W)	TIME OF LOCAL MIDNIGHT	OBSERVER'S LONGITUDE (W)	TIME OF LOCAL MIDNIGHT
67½°	23.30 EST	97½°	24.30 CST/ 23.30 MST
70	23.40 EST	100	23.40 MST
72½	23.50 EST	102½	23.50 MST
75	24.00 EST	105	24.00 MST
77½	24.10 EST	107½	24.10 MST
80	24.20 EST	110	24.20 MST
82½	24.30 EST/ 23.30 CST	112½	24.30 MST/ 23.30 PST
85	23.40 CST	115	23.40 PST
87½	23.50 CST	117½	23.50 PST
90	24.00 CST	120	24.00 PST
92½	24.10 CST	122½	24.10 PST
95	24.20 CST	125	24.20 PST

THE STAR MAPS

The accompanying eight maps show all the stars in the sky brighter than about mag. 5·2. Maps 1 and 8 show the north and south polar regions beyond a declination of 50°, while the intervening maps show strips of the sky centered on the celestial equator and extending to 60° and –60° in Dec. and for 5 hours in R.A.

The brightnesses of the stars are indicated, at magnitude intervals, by dot size. Identification letters or numbers have been applied only to those stars brighter than mag. 4·5 (some stars brighter than this limit have never been assigned symbols, and therefore appear unidentified), except where the object has been mentioned in the text.

Nebulae and clusters are indicated by a circle with the N.G.C. number alongside (no prefix), or the Messier number (preceded by an M). Binocular variable stars which can fall below the naked-eye limit at minimum are indicated by a small triangle.

Table 12 *The Greek Alphabet*

α	Alpha	ν	Nu
β	Beta	ξ	Xi
γ	Gamma	ο	Omicron
δ	Delta	π	Pi
ϵ	Epsilon	ρ	Rho
ζ	Zeta	σ	Sigma
η	Eta	τ	Tau
θ	Theta	υ	Upsilon
ι	Iota	φ	Phi
κ	Kappa	χ	Chi
λ	Lambda	ψ	Psi
μ	Mu	ω	Omega

☆ 7 ☆

Around the Constellations

Star Maps on pages 123–130

The night sky is so full of objects of interest and beauty to observe in a pair of binoculars or a very small telescope that the selection of a limited number for a list involves impossible problems of choice. All that can, in practice, be done is to select some of the more striking examples of colored stars, double stars, variable stars, nebulae and clusters, in the full knowledge that at least as many worthy objects have had to be omitted. The main criteria must be the allied ones of brightness and ease of location. From a practical aspect it may well be found difficult or impossible to locate a 6th magnitude star when even the main shapes of the constellations are, as yet, uncertainly traced; furthermore, the amateur living in a town may find such objects too faint in his brightly lit sky to be worth looking at. However, some faint but easy-to-find objects have been included, as tests of atmospheric transparency. Instrumental tests are best afforded by some close or very unequal double stars, and it is the writer's experience that many amateurs like to try their hand at difficult rather than easy objects, the satisfaction of the observation lying in the bare achievement rather than in the vividness or delicacy of the spectacle. The town-based observer will certainly find far more interest in the field of binocular doubles than in that of clusters and nebulae, which will, with some worthy and welcome exceptions, be mere stains with which he can do little but verify their existence.

The colors of stars show well only when they appear at the right brightness through the instrument concerned, although some vivid tints (relatively speaking) can be caught in much fainter objects. In general, comments on color are confined to the brightest stars in the constellation. Of nebulae and clusters, as has been mentioned, only the brightest are included. Careful sweeping through the Milky Way on a transparent night will reveal many more gleams and misty patches which can usually be verified from *Norton's Star Atlas* once they are found in the sky. All the objects mentioned in this chapter are marked on the maps, but the fainter ones can be sought more confidently with *Norton's Star Atlas,* which shows stars a magnitude fainter than the book maps, whose limit is the 5th magnitude, and offers more field stars to indicate the exact situation of small and elusive objects.

Variable stars offer special problems. Although many long-period variables can reach binocular brightness near maximum, it would be necessary to have a comprehensive chart of the area to ensure positive identification unless observations were protracted in order to identify the temporary intruder—which might, in any case, be invisible for months on end, if it happened to be near minimum. The proper course of events, if one wishes to observe these and similar stars, is to obtain charts from the appropriate source, the American Association of Variable Star Observers (AAVSO), to be found in Appendix 1. The only variables included in these notes are, therefore, the ones that can be identified from the maps, and whose variations can be watched adequately with binoculars; some special variable-star charts are given in the following chapter.

After each constellation is given the approximate date on which it is due south at 10 PM. This gives some guide to the best time for observation. The objects in each category are listed in order of right ascension, from west to east.

Andromeda (November 10; MAPS 2, 7)

A large but not easily identified mid-northern constellation; its three brightest stars are spread somewhat SW–NE to the south of Cassiopeia. The Great Square of Pegasus adjoins to the west.

Colored star α: mag. 2·1; a fine golden tint.

Double star 56: mags. 5·8, 6·1; 300°; 190″. This easy pair will be swept up without difficulty.

Nebula M.31 (N.G.C. 224): Known as the Great Nebula in Andromeda, this extended nebulous blur is a galaxy of the same order of size as the Milky Way (about 30,000 parsecs across), lying some 650,000 parsecs away—which means that the light we see left the galaxy over two million years ago. This object is bright enough to be seen easily with the naked eye on a clear night, even from well-lit suburban regions. Telescopically it is disappointing, for it remains a ghostly and almost featureless blur with all but the most powerful instruments, which begin to resolve its brightest stars. There is, however, a very real possibility of the modestly equipped amateur repeating the observation of a *supernova,* a catastrophic exploding star, that was made by amateurs in 1885. At maximum, this wrecked star shone out from near the nucleus at mag. 6—almost equal to the total brightness of the galaxy!—and faded away over a few weeks. Such phenomena probably occur at intervals of some hundreds of years, but one could be seen at any time.

Antlia, The Air Pump (March 25; MAP 4)
A faint mid-southern group adjoining the Milky Way in Pyxis and Vela. The brightest stars are only of the 4th magnitude.

Colored star α: mag. 4·2, reddish.

Double star ζ: an easy 6th magnitude pair.

Apus, The Bird of Paradise (June 20; MAP 8)
A far southern group, easily identified by its triangle of 4th magnitude stars near the pole that culminate on winter evenings.

Double star δ: a fine wide pair of 4th magnitude stars.

Aquarius, The Water Bearer (September 25; MAP 7)
A somewhat rambling equatorial constellation, with its brightest stars mostly huddled just south of Pegasus. The southern part of the group is very barren.

Double star τ: a very wide binocular pair, mags. 5 and 6; the brighter star yellow.

Cluster M.2 (N.G.C. 7089): a magnificent globular cluster about 6′ across, visible as an almost stellar spot in binoculars.

Aquila, The Eagle (August 15; MAPS 6, 7)
Aquila's leader α (Altair, mag. 0·7), forms the southern apex of the "summer triangle," the other corners being marked by Vega (α Lyrae) and Deneb (α Cygni). This huge triangle forms a useful reference in the summer sky for northern observers. Aquila itself lies on the celestial equator and contains rich Milky Way fields.

Colored star ϵ: compare the yellow tint of this mag. 4·0 star with its white mag. 3·0 companion ζ (it is instructive to remember that the difference in brightness between these stars is exactly one magnitude).

Double star: Despite its richness, there are few binocular pairs. Star 57 (mags. 5·8, 6·5; 170°; 36″) is a good test for a small instrument.

Variable star η: Cepheid type, mag. range 4·1–5·4, period 7·2 days. Therefore it periodically outshines its neighbor ι (mag. 4·4), but is usually the fainter of the two.

Ara, The Altar (July 10; MAPS 6, 8)
A fine southern Milky Way group, adjoining Scorpius.

Double star κ: a difficult test object for a large binocular, mags. 5·3, 9·7; 160°; 76″.

Clusters N.G.C. 6193: an open cluster, with a mag. 6 star involved. N.G.C. 6397: a globular cluster, appearing as a misty patch.

Aries, The Ram (November 30; MAP 2)
A conspicuous little north-equatorial group to the south of the Andromeda line, forming a roughly equilateral triangle with α and δ Andromedae

Double stars λ: mags. 4·9, 7·7; 46°; 37″. Beyond the reach of most 8 × 30 binoculars. Nearby is another interesting test object, star 14 (mag. 5·1), which has two companions: a mag. 7·7 star at 278°, 106″ and (much harder) a mag. 8·7 star at 36°, 93″.

Auriga, The Charioteer (January 20; MAP 3)
Auriga is easily identified by its leading star α (Capella, mag. 0·1), which shines in the mid-northern sky between Gemini and Perseus with a distinctive yellow hue. This fine star is in much the same physical state as the Sun. Lacking binocular doubles, it contains three fine clusters and some superb Milky Way fields.

Clusters M.38, 36 and 37 lie across the line joining θ Aurigae and β Tauri. All are bright open clusters, visible with the slightest optical aid; but they are quite different in character. With 12 × 40 glasses, M.38 appears large and dim, with bright stars resolved, while M.36 is small and bright, and appears completely resolved into stars. The stars in M.37 are so faint and close-packed that it appears nebulous.

Variable stars ϵ: an eclipsing variable of the "dark-eclipsing" type, in which one of the two components of the binary is larger and dimmer than the other, producing a relatively rapid diminution of light when the bright component passes behind it. The period of ϵ is 27·1 years, with a brightness range from mag. 3·0–4·0. Its nearby companion ζ is another dark-eclipsing variable with a range of 3·8–4·3 and a period of 972 days. Both of these unusually long periods imply well-separated stars moving relatively slowly in large orbits. ζ Aurigae is one of the most massive systems known; the large, dim component is some 20 times as massive as the Sun, with a diameter of about 180 million miles. In 1982, ϵ Aurigae began one of its slow eclipses. Observers wishing to estimate the current brightness of these stars could employ η Aurigae (mag. 3·2) and 58 Persei (mag. 4·3).

Boötes, The Herdsman (June 2; MAP 5)

With its leading orange star α (Arcturus, mag. −0·1), the mid-northern group of Boötes heralds the coming of the northern summer. There is always the comfort, when seeing Arcturus rising in the east in the late evening, that winter has not long to go; and by the time it disappears into the evening twilight, the long warm days have arrived. Easily found by continuing the curve of the "handle" in Ursa Major, Boötes looks rather like a kite, with Arcturus dangling from its tail.

Colored stars α: deep yellow. This is a giant star over a hundred times as luminous as our Sun, and about 20 million miles across. Its motion across our line of sight (or *proper motion*) is exceptionally fast, and relative to the field stars it is moving at a rate of 1° in 16 centuries.

Double stars ι: mags. 4·9, 7·5; 33°; $38\frac{1}{2}''$; a very hard test. δ: mags. 3·5, 8·7; 79°; 105″; primary yellowish. μ: mags. 4·5, 6·7; 171°; 108″; an easy pair.

Caelum, The Chisel (November 30; MAP 3)
A faint mid-southern group west of Columba, containing nothing of particular interest to the binocular observer.

Camelopardalis, The Giraffe (January 23; MAP 1)
An inconspicuous far northern constellation which follows Cassiopeia in its diurnal path around the pole star. This group, paradoxically, contains numerous fine telescopic double stars and an interesting close binocular pair.

Double star β: mags. 4·2, 8·8; 208°; 81″. The primary is the brightest star in the constellation, of a fine yellow color, and can be identified by a wide 6th mag. pair to the south in the same field. Probably too difficult for 8 × 30 binoculars.

Cancer, The Crab (February 28; MAP 4)
Rather inconspicuous but easily found, lying midway between Regulus in Leo and Castor and Pollux in Gemini. Its most striking feature is the naked-eye cluster Praesepe.

Double star ι: mags. 4·2, 6·6; 307°; 30½″. Very close with 12 × 40 binoculars. The companion appears blue, perhaps from contrast with the yellowish primary.

Clusters M.44 (N.G.C. 2632): Praesepe (the Beehive). Binoculars, because of their wide field of view, will give a better view of this bright and scattered cluster than a powerful astronomical telescope. Paradoxically, because of its prominence, it tends to be overlooked in the search for fainter objects; examine it for some minutes on a clear, dark night. M.67 (N.G.C. 2682): in the same binocular field as α, this cluster is much more remote than Praesepe and appears as a misty spot.

Canes Venatici, The Hunting Dogs (May 7; MAP 5)
This constellation contains only one prominent star, α, mag. 2·8 (Cor Caroli, or Charles' Heart), which is easily found beneath the canopy formed by the handle of Ursa Major.

Colored star Y: 5th mag., slightly variable star, which is one of the reddest known.

Cluster M.3 (N.G.C. 5272): a superb globular cluster in powerful telescopes, and easily found with binoculars as a hazy disk.

Nebula M.51 (N.G.C. 5194): faint in binoculars, appearing as an irregular, extended haze 3° southwest of the last star (η) in the handle of Ursa Major. This is a galactic system, the first ever to have its spiral nature observed visually, using Lord Rosse's huge 6-foot aperture reflecting telescope at Birr Castle, Co. Offaly, Eire, in 1845.

Canis Major, The Greater Dog (February 1; MAP 3)
A small bright constellation somewhat south of the equator, to the east of Orion. It is impossible to overlook since its leading star α (Sirius), at mag. –1·45, is the brightest star in the sky. The Milky Way touches its eastern border, and fine sweeping is to be had.

Sirius owes its brightness partly to its luminosity (27 times that of the Sun), and partly to its nearness (only $3\frac{1}{2}$ parsecs away). Its white dwarf companion, with a diameter near to that of the planet Uranus but a mass corresponding to that of the Sun, requires a moderate telescope for its detection even when favorably placed in its orbit.

Clusters M.41 (N.G.C. 2287): despite its low altitude, this is an easy binocular object for most observers, and a naked-eye object from lower latitudes. Two other open clusters are N.G.C. 2360 (faint) and N.G.C. 2362, involved with star τ.

Canis Minor, The Lesser Dog (February 15; MAP 3)
A small constellation distinguished by α (Procyon), which forms a triangle with Sirius and α Orionis. It lies to the north of the celestial equator.

Colored star α, mag. 0·35, has a distinctive mid-yellow tint.

Double star 14: a faint but interesting triple, quite easily located. A mag. 5·5 star has 8th and 9th magnitude companions at 82°, 95″ and 149°, 124″ respectively; the latter may just be glimpsed with adequate binoculars in a good sky.

Capricornus, The Sea-Goat (September 8; MAP 7)
A somewhat obscure south equatorial group, distinguished by its naked-eye double α south of θ Aquilae.

Double stars α: 3·6, 4·3, very wide, both yellow. δ: mags. 5·5, 9·0; 180°; 56″, a difficult test. β, mag. 3·1, has a wide 6th magnitude companion.

Cluster M.30 (N.G.C. 7099): a globular cluster revealed as a faint nebulosity $\frac{1}{2}°$ northwest of a 5th mag. star.

Carina, The Keel (March 2; MAPS 3, 4, 8)
This bright far southern constellation precedes the Southern Cross around the pole. It is a part of the once-extensive group of Argo, the Ship, and contains much of interest for binoculars and telescopes. Its leader α (Canopus), at mag. −0·7, is the second brightest star in the sky, and it lies near the same right ascension as Sirius.

Variable stars R: with mag. range 4–10, period 309 days, this is one of the brightest long-period variables in the sky, and is visible throughout its range with powerful binoculars, being a naked-eye object near maximum. η: one of the most remarkable stars in the sky. In the first half of the nineteenth century it fluctuated over a wide range of naked-eye brightness, almost rivaling Sirius in 1843–50. About 1864 it ceased to be visible with the naked eye, and is now of the 7th magnitude, slightly variable. Binoculars show it involved with the bright gaseous nebula N.G.C. 3372: a brilliant telescopic object.

Clusters N.G.C. 2516: a bright open cluster nearly 1° across, visible to the naked eye. N.G.C. 3114: a large cluster of faint stars. θ: a spectacular grouping of bright stars. N.G.C. 3532: a bright irregular open cluster.

Cassiopeia (November 9; MAPS 1, 2)
A far northern group, on the opposite side of the pole star from Ursa Major. Its "W" configuration is unmistakable. Cassiopeia lies in a rich part of the Milky Way, and the sweeping on an autumn evening is very fine.

Cassiopeia and Ursa Major between them constitute the Cosmic Clock (Fig. 24). Draw an imaginary line from β Cassiopeiae to γ Ursae Majoris, passing through the pole star. Taking β as the pointer on the hand, we have here a huge clock sweeping out the sidereal time in an anti-clockwise direction.

Colored star α: compare the yellow tint of this star (mag. 2·2) with that of the white β (mag. 2·3).

Double stars β: note two 7th mag. binocular pairs lying immediately to the south. α: 2·2, 9; 280°; 63″; probably beyond the range of normal binoculars.

Variable stars ρ: a naked-eye variable of unknown type; normally of the 4th magnitude, it has been known to fade to the 6th, and should be watched. A useful series of comparison stars are χ, 4·2; o, 4·5; λ, 4·75 and σ, 4·9. γ: a rapidly spinning star, subject to unpredictable brightenings at long intervals; normally about mag. 2·4, it attained 1·7 in 1937.

Clusters: There are many condensations in this part of the Milky Way. The fine cluster M.52 (N.G.C. 7654) is an easy binocular object. N.G.C. 7789, a large group of faint stars, is seen as a conspicuous glow under good conditions; N.G.C. 663, easily swept up near δ and ϵ, is an easy object.

Centaurus, The Centaur (April 30; MAPS 4, 5, 8)

A large mid-southern group to the north of Crux; but much of it contains only faint stars. Its leaders, α and β, form a conspicuous pair. α, mag. −0·3 (yellow), is the closest naked-eye star to the Sun, lying at a distance of only 1·3 parsecs.

Double star δ: mags. 2·9, 4·8; 325°; 69″; a nearby third star makes it triple.

Clusters N.G.C. 5045: an open cluster on the edge of the Coalsack, in Crux. N.G.C. 5139 (ω): glorious globular cluster easily visible with the naked eye, diameter 20′. Considered by many to be the finest stellar object in the sky. N.G.C. 5662: an open cluster with a mag. 7 star involved.

Cepheus (October 29; MAP 1)

A far northern autumn constellation. It is not easy to make out until the parallelogram ι, β, α, δ is established. It contains a number of interesting stars, but lacks clusters.

Colored star μ: variable, but about mag. $4\frac{1}{2}$. The "garnet star" of Sir William Herschel; a fine claret color even with binoculars, and it can be found by tint alone.

Double & variable star δ: the prototype of the pulsating Cepheid variables, mag. range 3·6–4·3, period 5·37 days. It therefore bridges the gap between ζ (mag. 3·4) and ϵ (mag. 4·2). A companion to δ, of mag. 5·3 at 192°, 41″ is an easy binocular object.

Cetus, The Whale (November 10; MAP 2)

An extensive equatorial constellation containing few prominent stars, marking a dull region of the sky south of Pisces. To mid-latitude observers, star β (mag. 2·0) fills a vacancy in the low southern sky on autumn evenings.

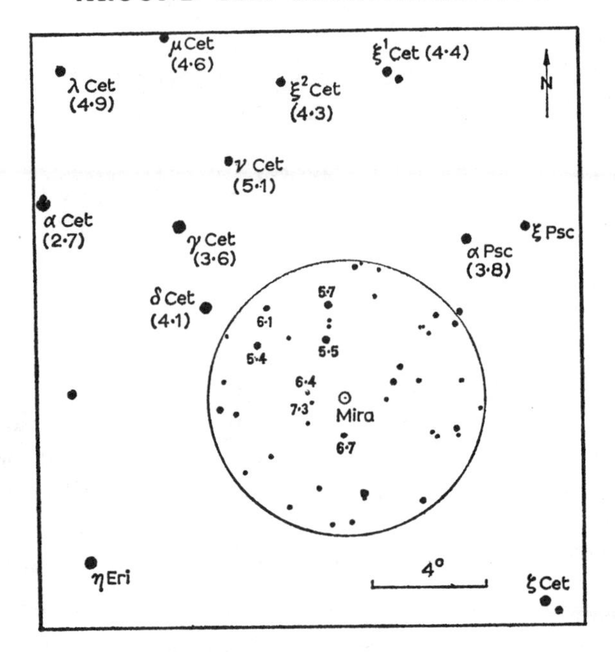

Fig. 25. *Chart for o Ceti (Mira)*

The outer part of the chart shows only the brighter naked-eye stars (comparison magnitudes given where appropriate). The 10° circle shows all stars down to magnitude 7·5, with comparison star magnitudes indicated.

Double star α: a fine wide binocular pair, mags. 2·5 (yellow) and 5·5.

Nebula M.77 (N.G.C. 1068): a spiral galaxy visible in good binoculars as a hazy spot. This object is well worth looking for, because it is the most remote of all the Messier objects (distance about 16 million parsecs, or over 20 times that of the Andromeda nebula), and represents, perhaps, the furthest reach of a pair of binoculars.

Variable star o: Mira, the "wonderful star." The brightest long-period variable in the sky at its extreme maximum (about mag. 2), but usually achieving mag. 3–4. With the aid of a chart (Fig. 25), it can be distinguished with binoculars at most minima (mag. 9–10). Mira is by no means as red as many stars of its class, and binoculars will show no more than a yellow tint. An 8th mag. companion lies within about 2′ of the variable.

Chamaeleon (March 31; MAP 8)
A small far southern group, very near the celestial pole.
Double star δ: mags. 4·6 and 5·5, very wide.

Circinus, The Compasses (May 29; MAP 8)
A far southern group involved in the Milky Way near α Centauri. Fine sweeping, but much of the galaxy here is obscured by dark matter.

Columba, The Dove (January 16; MAP 3)
A small but easily distinguished mid-southern group south of Lepus. Although lacking in striking individual objects, its brighter stars form, to observers in high northern latitudes, interesting tests of southern visibility at the time when Orion is on the meridian. α (mag. 2·6) lies at declination 34° 06′ south, and β (mag. 3·1) at 35° 47′ south.

Cluster N.G.C. 1851: a small globular cluster, appearing as a nebulous patch.

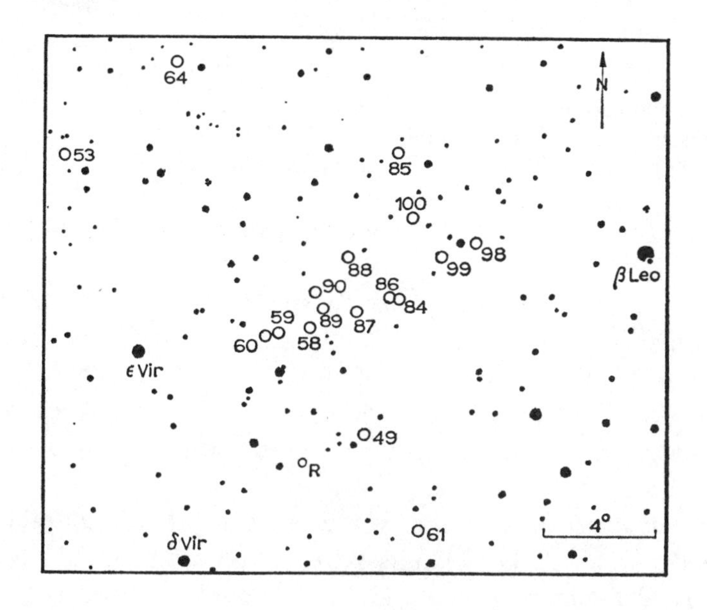

Fig. 26. *Chart for the Coma-Virgo galaxies*

Open circles indicate the positions of the Messier Objects in this region, with the number alongside. Some, such as M.98, are too faint for binocular observation. Note that M.53 is a globular cluster, not a galaxy.

Coma Berenices, Berenice's Hair (May 2; MAP 5)
This faint mid-northern group lies roughly in the triangle formed by α Canum Venaticorum, Arcturus, and β Leonis. It could really be called an extensive star cluster, particularly in the region around γ, which presents a fine spectacle in binoculars. Coma and its southern neighbor Virgo contain the most extensive grouping of external galaxies to be found in the sky, but relatively few are bright enough to be visible with binoculars.

Double stars 12: mags. 4·8, 8·3; 167°; 65″, and 17: mags. 5·4, 6·7; 251°; 145″, are easily found in the main "cluster."

Cluster M.53 (N.G.C. 5024): a bright globular cluster near α, an easier object than any of the nebulae in the region.

Nebulae M.85 (N.G.C. 4382): dim but large in 40mm binoculars. M.64 (N.G.C. 4826): easily found northeast of a mag. 5 star, a small and bright nebulosity with a central condensation. Figure 26 shows, on a larger scale, the positions of these and other galaxies in the Coma-Virgo grouping.

Corona Austrina, The Southern Crown (July 30; MAP 6)
A mid-southern group south of Sagittarius. There are no stars brighter than the 4th magnitude.

Double star κ: a bright and wide triple.

Cluster N.G.C. 6541: a fairly bright, condensed globular cluster.

Corona Borealis, The Northern Crown (June 19; MAP 5)
A distinctive little constellation immediately to the east of Boötes, in mid-northern latitudes, consisting of a semicircle of stars embracing α (mag. 2·3).

Double star ν: a very wide pair, mags. 4·8 and 5·0.

Variable stars R: normally visible as a mag. 6 star inside the "crown," this star occasionally falls, in a week or two, to a precipitous and usually short-lived minimum between the 11th and 14th magnitudes. It may, however, be months or years before the star once more regains its normal brightness. The onset of a fall can be best detected with binoculars, and a chart of the region is given (Fig. 27). T: a novalike variable, normally of the 10th magnitude, which rose to the 2nd magnitude in 1866 and to the 3rd in 1946. These two stars in Corona Borealis, being so unpredictable, should be checked on every suitable night.

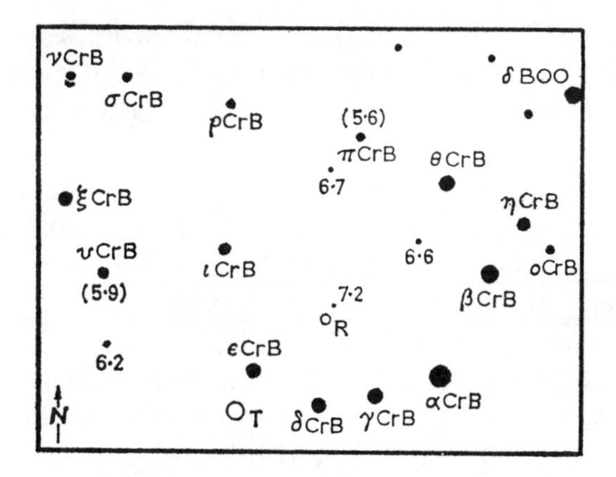

Fig. 27. *Chart for R Coronae Borealis*

Comparison star magnitudes are indicated.

Corvus and Crater, The Crow and The Cup (April 20; MAPS 4, 5)
Two small equatorial groups south of Virgo. The trapezium of Corvus is distinctive, but neither contain any remarkable binocular objects.

Crux, The Cross (April 27; MAP 8)
A brilliant, compact far southern group, lying in a prominent region of the Milky Way against which the dark obscuring cloud known as the Coalsack (7° × 4°) is strongly outlined.

Colored star: Compare the yellow tint of γ (mag. 1·7) with the white star β (mag. 1·1).

Double stars α: a fine binocular pair, mags. 1·6, 5·1; 202°; 90″ β and γ also have wide binocular companions.

Cluster N.G.C. 4755: a superb, bright open cluster surrounding the star κ, with a diameter of about 10′.

Cygnus, The Swan (August 20; MAPS 6, 7)
This mid-northern group offers the finest star fields to be found in the northern sky, although they are much inferior to those of Scorpius and Sagittarius seen from lower latitudes. The cruciform shape of this large and distinctive constellation has earned it the common nickname of the Northern Cross. Its leader α (Deneb, mag. 1·25), forms part of the great summer triangle (page 95).

Double stars β: mags. 3·2, 5·4; 54°; 34″. Yellow and blue, a fine pair in a small telescope and just divided with good binoculars. 16: mags. 6·3, 6·4; 134°; 39″; a close, yellowish binocular pair. o: a mag. 4·0 yellow star showing brilliant contrast with its wide, blue mag. 5·0 companion. Note also the fainter companion to the brighter star: mag. 6·9; 173°; 107″. The main pair is one of the most colorful binocular doubles of all. γ: users of large binoculars might search for the mag. 9½ companion at 196°, 141″; the main star, mag. 2·2, lies in a fine part of the Milky Way. Many other moderately bright pairs may be swept up in this region.

Variable stars χ: a long-period variable which occasionally reaches the 4th magnitude; it can fade to the 14th at minimum. The period is about 406 days, and a chart of the field is given. P: a curious star which rose to the 3rd magnitude in 1600 and is now about mag. 5 and slightly variable. This is one of the most luminous known stars in our galaxy, equivalent in brightness to about 150,000 Suns!

Nebulae: Two very large and dim nebulae form excellent tests, but should be attempted only on the darkest nights and away from artificial lights. N.G.C. 6992: the brightest part of an incomplete circle of nebulosity involving and to the east of star 52. This, the brightest arc, is about 2° ENE of the star. N.G.C. 7000: 7 × 50 or other large-aperture binoculars show a glow, larger than the Moon, northwest of star ξ; this is the center of an extensive nebulosity to be seen only on photographs, and, because of its shape, it is frequently referred to as the North America Nebula.

Cluster M.39 (N.G.C. 7092): a coarse triangular cluster of mag. 7–8 stars.

Delphinus, The Dolphin (August 31; MAP 7)
A small but very distinctive north equatorial group to the east of Aquila. Its leader α (mag. 3·9) has a wide binocular companion. It became famous in 1967 through the discovery within its precincts of an unusual slow nova by British amateur G. E. D. Alcock.

Dorado, The Swordfish (January 16; MAPS 3, 8)
In itself this is a faint and rambling mid-southern group, but it contains most of the remarkable Greater Magellanic Cloud, a companion galaxy to our own.

Variable star β: Cepheid type, mag. range 3·8–4·8, period 9·8 days.

Cluster & Nebula N.G.C. 2027: in the Cloud, a cluster partly resolved with binoculars. Further south is N.G.C. 2070, a brilliant nebulosity surrounding the star 30 and visible to the naked eye. There are many other groups in this sparkling area. The Greater Cloud has a diameter of about $\frac{2}{5}$ that of our own galaxy, and lies some 50,000 parsecs away.

Draco, The Dragon (May 25; MAPS 1, 6)
A far northern constellation which winds its way for almost 180° around the celestial pole, so that part at least is well placed for observation at some time on any night. It contains some fine binocular doubles, but identification is not always easy to make.

Double stars ν: mags. 5·0, 5·0; 312°; 62″. Both stars are white, and form one of the finest binocular pairs; well seen with 8 × 30 glasses, but more impressive with a higher magnification. ψ: mags. 4·9, 6·1; 15°; 30″; a very neat pair. 39: this mag. 5·0 star has a 7th magnitude companion, wide. o: mags. 4·9, 7·9; 322°; 35″; a difficult pair.

Equuleus, The Little Horse (September 8; MAP 7)
A very small group, lying near the equator between Delphinus and Pegasus. It contains only three mag. 4 stars.

Double star γ: 4·5, 6·0; a wide binocular pair.

Eridanus, The River (December 10; MAPS 2, 3)
The river flows from its leader α (Achernar, mag. 0·5), in a barren part of the far southern sky, to its extensive delta west of Orion and Lepus. The star γ is the only one likely to be noticed at a casual glance by north temperate observers.

Colored star γ: mag. 2·9, distinctive red tint.

Double star o^2: mags. 4·5, 9·7; 105°; 83″, primary yellow. A difficult test object for large binoculars.

Fornax, The Furnace (December 2; MAP 2)
A mid-southern group in a dull region between Eridanus and Sculptor. Its brightest star is only of the 4th magnitude, and it offers little to the binocular observer.

Gemini, The Twins (February 5; MAP 3)
A fine mid-northern group in the form of a long parallelogram,

its western end meeting Orion in the Milky Way, where we find some fine fields. It is easily identified from its leading stars α (Castor, mag. 1·6) and β (Pollux, mag. 1·1). These stars form one of the many instances where the original lettering was clearly not in the order of brightness.

Colored stars β: compare the rich yellow of this star with the very pale tint of α. μ: mag. 2·9, orange. η: mag. 3·5, orange, and slightly variable.

Double stars ν: this mag. 4·1 star has a 9th magnitude companion at 329°, 113″. ε: mags. 3·2, 9·2; 94°; 110″. A hard test, seen with averted vision only, 12 × 40 binoculars. ζ: mags. 3·7–4·2, 8·0; 349°; 97″; a fine binocular pair. β has a 10th magnitude companion at 72°, 188″: an interesting test for a dark sky, keen eyes and large binoculars.

Cluster M.35 (N.G.C. 2168): one of the brightest groups in the northern sky; on a clear night it can be seen with the naked eye. Partly resolved with binoculars.

Grus, The Crane (September 28; MAP 7)
A small mid-southern group, containing some prominent stars.

Colored stars: Compare the tints of β (mag. 2·1, orange) and α (mag. 1·7, white).

Double stars δ and μ are both very wide pairs.

Hercules (July 13; MAP 6)
Large but unremarkable to the naked eye, this mid-northern constellation lies in the large area marked out by Corona Borealis, Vega and α Ophiuchi. The most distinctive feature of the group is the quadrilateral formed by ε, ζ, η and π. Hercules contains a great many double stars for the telescopic observer.

Colored star α: a fine red star of about mag. 3½ (variable; see below).

Double stars κ: not too easy to identify, but look between γ Herculis and γ Serpentis: mags. 5·3, 6·5; 12°; 28″; neatly split with 12 × 40 glasses. γ: mags. 3·8, 9·8; 230°; 42″; probably beyond the range of all but giant binoculars.

Variable star α: mag. range 3·0–4·0, irregular. Suitable comparison stars are δ (3·1), κ Ophiuchi (3·2), and γ (3·8).

Clusters M.13 (N.G.C. 6205): the Great Cluster in Hercules, one of the few globular clusters in the sky that are visible to the naked eye under good conditions. It appears as a circular misty patch in binoculars. Another fine globular in the same constellation, M.92 (N.G.C. 6341), can also be found quite easily, and shows a more condensed nucleus.

Horologium, The Clock (December 10; MAPS 2, 8)
A very obscure southern group, lacking interesting binocular objects.

Hydra, The Water Snake (April 15; MAPS 4, 5)
The western end of this long south-equatorial group is distinctively marked by the yellow α (mag. 2·0), which shines in the empty sky south of Leo's paws; and by the grouping between Regulus and Procyon. The rest of the constellation is very obscure.

Variable stars R: mag. range 3–11, period 388 days. This is not too difficult to pick up when bright, because it lies about 2° east of γ, the brightest star in the region south of Spica. U: mag. range 4·5–6, period 450 days (with large variations); can be identified by its rich red color, forming a triangle with μ and ν.

Hydrus, The Water Snake (November 25; MAP 8)
This south circumpolar serpent is more compact than its equatorial counterpart, but contains little in the way of interesting objects. Note β, a wide pair.

Indus, The Indian (September 11; MAPS 7, 8)
A faint mid-southern group, of limited interest to binocular users.

Lacerta, The Lizard (September 25; MAP 7)
This is a small mid-northern constellation and contains no bright stars, but it nudges a fine section of the Milky Way between Cassiopeia and Cygnus. Several binocular pairs will be found by sweeping. Despite its small size, three naked-eye novae have occurred within its limits during the present century, so it is worth being on the alert when viewing the region.

Leo, The Lion (April 1; MAP 4)
A large and obvious northern equatorial group, and remarkable in having a definite resemblance to its namesake; the mane and general form of a lion looking westwards is very clear. Its leading star α (Regulus, mag. 1·4) lies almost on the ecliptic, and can occasionally be occulted by the Moon or a planet.

Leo is one of the galaxy-rich areas of the sky.

Double stars α: mags. 1·4, 7·6; 307°; 177″. τ: mags. 5·2, 8·1; 178°; 90″. A very neat pair, the brightest in a line of stars. 93: mags. 4·5, 8·6; 355°; 74″; difficult.

Nebulae M.65 (N.G.C. 3623) and M.66 (N.G.C. 3627): two close galaxies, visible in binoculars as a somewhat extended nebulous patch; M.66, to the southeast, is the brighter of the two.

Leo Minor, The Lesser Lion (March 23; MAP 4)
A small group over the head of Leo, lacking any interesting binocular objects.

Lepus, The Hare (January 15; MAP 3)
A small, bright constellation, easily found beneath Orion's feet.

Double star γ: mags. 3·8, 6·4; 350°; 96″.

Cluster M.79 (N.G.C. 1904): a small but bright globular cluster, visible in binoculars as a hazy spot.

Variable star R: mag. range 5·5–10·5, period 432 days. This long-period variable may be recognized by its color should it be near maximum. It is one of the reddest stars known, but a telescope is required to show the tint well.

Libra, The Scales (June 10; MAP 5)
An extensive but faintly marked south-equatorial constellation lying between Scorpius and Virgo.

Double star α: this star, mag. 2·7, has a wide 6th magnitude companion.

Variable star δ: mag. range 4·9–6·0, period 2·3 days. A dark-eclipsing variable, not difficult to locate.

Lupus, The Wolf (June 8; MAP 5)
A small but bright mid-southern constellation, bordering the Milky Way in Scorpius and containing fine sweeping.

Double star η: mags. 3·6, 9·3; 248°; 115″.

Cluster N.G.C. 5822: a fine open cluster, diameter about 40′. N.G.C. 5986: a globular cluster only faintly shown with binoculars.

Lynx (February 19; MAPS 1, 3, 4)
The story goes that Lynx was so called because it requires a lynx-eyed person to distinguish it! Certainly this mid-northern

group occupies a great vacancy between Ursa Major, Auriga, and Cancer, and it contains nothing of interest to the binocular observer.

Lyra, The Lyre (August 1; MAP 6)
Small, but for its size one of the brightest constellations, especially as α (Vega, mag. 0·0) is the most brilliant star in the late northern summer sky. It passes almost overhead from mid-northern latitudes, and forms the western corner of the summer triangle.

Double stars ϵ: a wide pair of mag. $4\frac{1}{2}$ white stars, the southern being slightly the brighter; each of these is a telescopic double. ζ: mags. 4·3, 5·9; 150°; 44″. A very fine binocular pair. β: mags. 3·3–4·2, 8·6; 149°; 46″. Another attractive pair, the primary being the prototype of the "bright-eclipsing" variable stars (see below). δ: a naked-eye double, but binoculars show the fine contrast of yellow and white. θ: mags. 4·5, 9·2; 71°; 100″; an interesting light-test.

Variable star β: mag. range 3·3–4·2, period 12·9 days. Because the two components of the binary are of near equal brightness, a secondary minimum of mag. 3·8 occurs halfway through the cycle when the fainter star is occulted by the brighter; this minimum is negligible in the case of Algol-type (dark-eclipsing) stars, where one is very much brighter than the other. γ (mag. 3·2) and κ (mag. 4·3) form useful comparison stars.

Mensa, The Table (January 13; MAP 8)
A very dim south circumpolar group containing no star brighter than the 5th magnitude. A part of the Greater Magellanic Cloud encroaches on its northern border.

Microscopium, The Microscope (September 3; MAP 7)
A small mid-southern group east of Sagittarius, containing no objects of particular interest for binocular observers.

Monoceros, The Unicorn (February 5; MAP 3)
This visually inconspicuous equatorial constellation contains no star brighter than the 4th magnitude, but it lies in a marvelous part of the Milky Way, east of Orion, and contains some fine objects.

Clusters N.G.C. 2232: a bright, scattered group around a mag. 6 star. N.G.C. 2244: a magnificent binocular cluster, visible with the naked eye as a condensation in the Milky Way. M.50 (N.G.C.

2323): a condensed cluster appearing nebulous and somewhat elongated with binoculars.

Musca, The Fly (April 30; MAP 8)
A small far southern Milky Way group, south of Crux, containing some fine sweeping.

Cluster N.G.C. 4833: a globular cluster near δ.

Norma, The Square (June 18; MAPS 5, 6, 8)
A mid-southern Milky Way group south of Scorpius, containing some lovely fields.

Clusters N.G.C. 6067: an open cluster of faint stars, about 20′ across. N.G.C. 6087: a fine open cluster, with many stars resolved in binoculars. This cluster contains the Cepheid variable S, mag. range 6·1–6·8, period 9·75 days.

Octans, The Octant (MAP 8)
The south circumpolar constellation. A faint group; the closest star to the pole is the 5th-magnitude σ, about 1° away.

Ophiuchus, The Serpent-Bearer (July 10; MAP 6)
This large equatorial constellation, which is marked principally by boundary stars, extends from Hercules to Scorpius, where it enters one of the richest parts of the whole Milky Way. Here are some superb objects for the more powerful telescope, but many are poorly seen from more northern latitudes.

Double stars 53: mags. 5·8, 8·5; 191°; 41″. 67: mags. 3·9, 8·5; 143°; 55″.

Variable stars χ: irregular variable, mag. range 4·2–5·0; use φ (mag. 4·3) and ρ (mag. 4·6). 66: a "flare star," which occasionally undergoes very rapid brightenings—in just a few minutes—from its normal 5th magnitude appearance to perhaps mag. 3 or even brighter. Such rises may occur at intervals of weeks or months, and few have been observed.

Clusters N.G.C. 6171: a faint cluster, just glimpsed with binoculars. M.12 (N.G.C. 6218): a condensed open cluster, seen as a nebulosity. M.19 (N.G.C. 6273): a condensed open cluster. M.10 (N.G.C. 6254): another condensed cluster, appearing nebulous. N.G.C. 6633: a fine open cluster, just detectable with the naked eye. Note also the fine group of mag. 8 stars about 1° northeast of β. There are many condensations and groups in this wonderful region of the Milky Way.

Orion (January 15; MAP 3)
This undoubtedly is the grandest constellation of all, containing no less than seven stars brighter than the 2nd magnitude. It is fitting that this group should be placed on the celestial equator (the westerly belt star δ has a declination of $-0° 20'$), so that it can be well seen from all parts of the globe. The center of Orion, with its famous nebula, marks a relatively nearby region of recently born, hot young stars about 460 parsecs away, and these stars are all white or blue-white. To be included among these very young stars is λ (mag. 3·4), which is thought to have started shining not more than ten thousand years ago; it is probably the youngest naked-eye star in the sky.

Colored star α (Betelgeuse): rich orange, somewhat variable (mag. 0·4–1·3), but in an irregular manner; period 5–6 years? It makes a vivid comparison with the pure white β (mag. 0·15), the brightest star in the constellation.

Double stars δ: mags. 2·5, 6·6; 359°; 53″. An interesting and easily identified test for normal binoculars. σ: this very fine telescopic multiple star will show only one companion in binoculars: mags. 3·8, 6·5; 61°; 42″.

Nebula M.42 (N.G.C. 1976): the Great Nebula in Orion, an easy naked-eye object south of the belt stars δ, ϵ, and ζ. Patient watching with binoculars on a dark night shows considerable irregular extension to what, at first sight, appears a featureless nebulosity. At the center of the nebula, above the tip of a dark intrusion from the east, lies the star θ, a telescopically quadruple object. The diameter of the visible nebulosity is about 12 parsecs.

Pavo, The Peacock (August 14; MAP 8)
A far southern group, containing few bright stars.

Variable stars λ: mag. range 3·4–4·3, irregular. κ: mag. range 3·9–4·8, period 9·1 days. This is one of the brightest Cepheid variables in the sky.

Cluster N.G.C. 6752: one of the largest globular clusters in the sky, over half the Moon's apparent diameter in extent; a notable binocular object.

Pegasus (October 1; MAP 7)
Heralder of the northern autumn is the Great Square of Pegasus, a huge north-equatorial rectangle formed by α, β, and γ Pegasi

and α Andromedae. The area of Pegasus, however, spreads much further west, to the Cygnus region of the Milky Way.

Colored star β shows a marked yellow tint compared with the white α (mag. 2·5), and is supposed to be slightly variable, although it does not depart far from its mean magnitude (also 2·5). Systematic comparisons against α, γ (mag. 2·9) and α Andromedae (mag. 2·1) could be of value.

Double stars ϵ: mags. 2·5, 8·5; 320°; 143″. π: a mag. 4·5 star with a very wide mag. 6 companion.

Cluster M.15 (N.G.C. 7078): a bright globular cluster, well seen with binoculars as a circular nebulous patch.

Perseus (December 10; MAPS 2, 3)
Here, east of Cassiopeia, are some fine mid-northern Milky Way fields; and α (mag. 1·8) lies in a splendid stream of bright stars.

Colored star η (mag. 3·8) is a fine yellow star.

Double star 57: mags. 6·1, 6·8; 198°; 116″. A pretty pair, the brighter star yellowish.

Variable stars β: mag. range 2·1–3·4, period 2·87 days. The prototype of the dark-eclipsing binary variables. For most of the time the star shines at maximum brightness, then dims down in five hours as the large, dim component occults the smaller and brighter one. The original brightness is gained in another five hours. At maximum it matches γ Andromedae (mag. 2·1), while at minimum it is, for an hour or so, fainter than δ Persei (mag. 3·0). ρ: an irregular variable, mag. range 3·3–4·0, with a very approximate period of 40 days. Observation is hampered by the lack of suitable comparison stars; κ (mag. 3·8), 41 Arietis (mag. 3·6) and ϵ Persei (mag. 2·9) are probably the most suitable.

Clusters N.G.C. 869 & 884, also known as *h* and χ Persei, form the famous Double Cluster, visible with the naked eye on most nights as a condensation in the Milky Way between α Persei and δ Cassiopeiae. These open clusters of bright stars, each larger than the apparent diameter of the moon, form a splendid binocular spectacle; the region of the Milky Way in their vicinity is very fine. M.34 (N.G.C. 1039): another naked-eye open cluster, a fine sight in binoculars.

Phoenix (November 3: MAPS 2, 7)
A barren mid-southern group between Eridanus and Grus.

Pictor, The Painter (January 15; MAP 3)
A faint mid-southern group to the west of Canopus.

Pisces, The Fishes (October 25; MAPS 2, 7)
A large but very obscure north-equatorial constellation containing no stars brighter than the 4th magnitude; most of it lies in the region south of Andromeda and Pegasus.

Double stars φ^1: mags. 5·6, 5·8; 150°; 30″. A useful resolving test for common binoculars. ρ: a very wide 5th magnitude pair. κ: another wide pair of 5th and 6th magnitude stars.

Nebula M.74 (N.G.C. 628): a very faint binocular object, this spiral galaxy should be visible as a dim nebulosity in large glasses under good conditions. Location is not difficult, about 1½° ENE of η (mag. 3·6).

Piscis Austrinus, The Southern Fish (September 24; MAP 7)
This mid-southern group is marked to the naked eye by its leader α (Fomalhaut, mag. 1·2), a fine white star shining in a lonely part of the sky; but it contains little for the binocular observer. To observers in the northern states, Fomalhaut is the most southerly star likely to be noticed at a casual glance, its declination being −29° 50′.

Puppis, The Poop (February 7; MAP 3)
A part of the sundered ship Argo, this mid-southern constellation lies in a dense region of the Milky Way east of Canis Major, and extends almost to Canopus.

Colored star ζ: mag. 2·25, one of the hottest and bluest stars known.

Double stars π (yellow) marks a splendid group of four stars. Both κ and ξ (yellow) have wide binocular companions.

Variable star L_2: mag. range 2·6–6·0, a red star with a period of about 140 days.

Clusters There are many clusters in this part of the Milky Way. N.G.C. 2422: a brilliant open cluster, diameter 30′, visible with the naked eye. M.46 (N.G.C. 2437): a large, faint grouping in binoculars. M.93 (N.G.C. 2447): a bright cluster, diameter 25′, with stars resolved. N.G.C. 2451: a prominent group containing a 4th magnitude star, with, to the southeast, N.G.C. 2477, a large cluster of faint stars unresolved with binoculars.

Pyxis, The Compass (March 6; MAP 4)
Yet another fragment of Argo, this mid-southern group lies to the east of Puppis. Although small and poorly marked, its western border contains some good Milky Way sweeping.

Reticulum, The Net (December 18; MAP 8)
A small far southern group, containing little of interest.

Double star ζ: a wide binocular 5th magnitude pair.

Sagitta, The Arrow (August 15; MAP 6)
This small mid-northern group, lying north of Aquila, is inconspicuous to the naked eye, but its four brightest stars form a distinctive sight in binoculars. Fine Milky Way fields can be found here.

Double stars ϵ: mags. 5·7, 8·0; 81°; 89″; the brighter star yellow. θ: mags. 6·3, 7·3; 223°; 85″.

Cluster M.71 (N.G.C. 6838): a very compressed cluster, appearing as a faint nebulosity in binoculars.

Sagittarius, The Archer (August 10; MAP 6)
This noble mid-southern constellation, with its neighbor Scorpius and the southern tip of Ophiuchus, lies in the direction of the galactic center, and here we see the greatest depth of stars in the whole circle of the Milky Way. To observers in southern and low northern latitudes the fields here, whether binocular or telescopic, are granular with stars, and more pairs, groupings, open and unresolved clusters and nebulosities are visible than could be found in any catalogue. The Milky Way in this region is so brilliant that some of the star clouds are readily visible as hazy patches in a twilight sky strong enough to obscure all but the brighter stars. Only an outline of interesting objects can be given; and, of these, most are either invisible or but poorly seen from latitudes north of New York.

Variable stars X: mag. range 4·8–5·6, period 7·0 days; a Cepheid variable. The direction of the galactic center lies about 1° to the south of this star. W: mag. range 4·8–5·8, period 7·6 days; another Cepheid, easily found about 1° north of γ.

Clusters M.23 (N.G.C. 6494): a fine open cluster with a mag. 7 star close by; partly resolved in binoculars. M.8 (N.G.C. 6523): a coarse group of stars with associated nebulosity; a remarkable telescopic object. M.17 (N.G.C. 6618): the famous "Omega Nebula," visible as an irregular nebulosity, with a cluster, M.18 (N.G.C. 6613), about 1° to the south, appearing as a hazy patch in binoculars. Between this group and the star μ the naked eye perceives a great cloud (M.24) which binoculars show as a magnificent mass of stars. M.28 (N.G.C. 6626): a bright globular cluster. M.25 (I.C. 4725): a bright open cluster containing an 8th magnitude star. M.22 (N.G.C. 6656): a brilliant globular cluster, visible with the naked eye. M.55 (N.G.C. 6809): a large, bright globular cluster, much less condensed towards the center than most, appearing nebulous. This list by no means exhausts the clusters and nebulae that are worth observing.

Scorpius, The Scorpion (July 3; MAP 6)
A most distinctive constellation when seen to its full extent, led by the red giant star α (Antares); but the tail does not rise above the horizon in the northern states.

Colored star α: one of the reddest of the bright stars, and slowly variable (mag. range 0·9–1·8, period about 5 years). This is one of the largest stars known, with a diameter corresponding to about four times that of the Earth's orbit; but since the mass approximates to that of the Sun, the matter in the outer reaches of the star is extremely rarefied.

Double stars ω: a very wide pair of mag. 4 stars. ν: mags. 4·4, 6·5; 337°; 41″.

Clusters M.80 (N.G.C. 6093): a small, bright, condensed globular cluster. M.4 (N.G.C. 6121): a large, bright globular cluster some 2° west of Antares; a fine binocular object. N.G.C. 6231: a bright open cluster containing seven 8th magnitude stars, with fainter members. The region immediately north of this, towards μ, appears nebulous to the naked eye, and the view in binoculars or a small telescope can only be termed fabulous. M.6 (N.G.C. 6405): a wonderful open cluster, visible with the naked eye, looking, in binoculars, like a butterfly brooch. M.7 (N.G.C. 6475): another wonderful naked-eye cluster. Magnificent sweeping from here towards γ Sagittarii.

Sculptor (October 25; MAPS 2, 7)
An inconspicuous mid-southern group containing no star brighter than the 4th magnitude; it occupies a large vacancy to the east of

Fomalhaut and south of β Ceti. It contains a number of galaxies, and at least two are bright enough to be detectable with binoculars, though the lack of guide stars hinders the search.

Nebulae N.G.C. 55: a large nebulosity, elongated east-west. N.G.C. 253: a prominent nebulosity, considerably extended northeast to southwest.

Scutum, The Shield (August 1; MAP 6)
This south-equatorial asterism contains only faint naked-eye stars, but it lies in the same galactic region as Sagittarius and Scorpius, and offers probably the finest Milky Way sweeping from high northern latitudes. On a clear night, the naked eye will show much local "patchiness" due to star clouds and dark nebulae. Scutum lies in one of the nova-prone regions of the sky, and 7th magnitude novae were detected within its confines in 1970 and 1975.

Colored star α: mag. 3·85, yellow.

Variable star R: mag. range 4·8–6·5, period about 140 days, with irregularities. This star is easily located near β, and a chart is given in Figure 33.

Clusters M.26 (N.G.C. 6694): only faintly seen in binoculars. M.11 (N.G.C. 6705): a magnificent, irregular telescopic open cluster, but well seen even with binoculars, which begin to resolve it; it lies at the nucleus of a superb Milky Way region.

Serpens, The Serpent (July 10; MAPS 5, 6)
This equatorial constellation is divided into two separate minor figures: Serpens Cauda (the body) and Serpens Caput (the head) on the east and west flanks of Ophiuchus; but the stars are labeled as though belonging to a single group.

Cluster M.5 (N.G.C. 5904): a magnificent globular cluster, well seen in binoculars, lying near a 5th magnitude star.

Sextans, The Sextant (March 22; MAP 4)
A faint asterism beneath Leo's paws, containing nothing of binocular interest.

Taurus, The Bull (January 1; MAPS 2, 3)
A large north-equatorial constellation containing mostly faint stars, but immediately identifiable by its fine reddish α (Aldebaran, mag. 0·9) and the associated Hyades cluster.

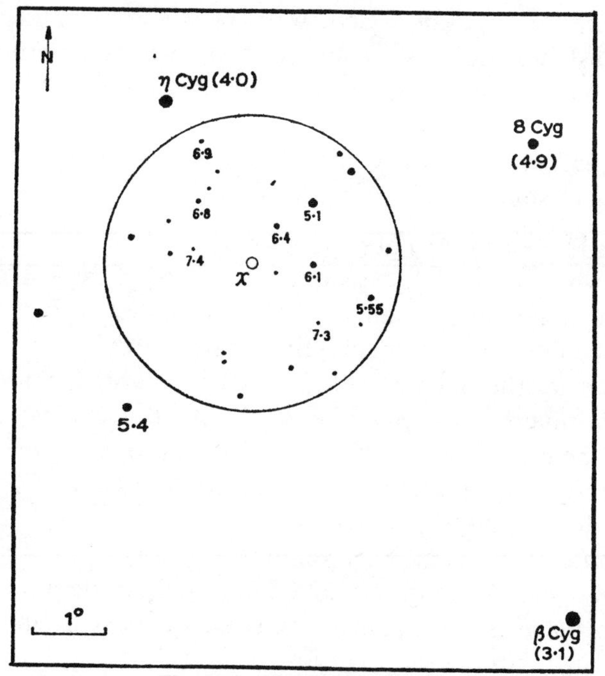

Fig. 28. *Chart for χ Cygni*

The 4° circle shows all stars down to magnitude 7·5, with comparison star magnitudes indicated.

Colored star α: a rich orange, intermediate in tint between Capella and Antares.

Double stars φ: mags. 5·1, 8·5; 255°; 51″; primary yellow. τ: mags. 4·3, 8·6; 213°; 63″.

Clusters M.45: the Pleiades, the most famous group in the sky. A cluster of young stars no more than 130 parsecs away, and possibly only 100 million years old. These stars form a convenient test of naked-eye acuity. The six brightest are easily seen by most eyes, but there are 11 stars in the group brighter than mag. 6, and some sharp-eyed observers have seen 14 or more. The accompanying figure gives a chart of the group. Because of the wide extent of the cluster, binoculars will give a more satisfactory view than can a more powerful instrument of restricted field.

There are a number of faint nebulous patches in the Pleiades, representing the remaining interstellar matter from which the cluster stars were formed. The brightest (N.G.C. 1435) extends for nearly ½° in a southerly direction from Merope, the southernmost bright star in the group, and has frequently been seen using 10 × 50 binoculars. Sky conditions must, however, be first-class for such an observation to be possible.

The Hyades cluster, to the west of Aldebaran, is very bright and scattered and fills a binocular field. The Hyades are three times closer to the Sun than are the Pleiades, and Aldebaran is much closer still, not being involved with the cluster in any way.

Nebula M.1 (N.G.C. 1952): the famous Crab Nebula, an object of almost continuous attention by professional astronomers, may just be glimpsed as a faint, hazy, almost stellar spot about 1° northwest of ζ. The remains of a supernova observed in the year 1054, this tiny smudge represents the site of one of the most titanic upheavals known in the natural world.

Telescopium, The Telescope (August 9; MAP 6)
A faint mid-southern group to the east of Ara.

Double star δ: a wide 5th magnitude pair forming an attractive group with α.

Triangulum, The Triangle (November 25; MAP 2)
A small, faint mid-northern group south of γ Andromedae; but the triangle is plainly marked by three moderate stars.

Nebula M.33 (N.G.C. 598): one of the most challenging binocular objects in the sky, although unmistakable once found. This is a large spiral galaxy, almost face-on and nearly ½° across, but of very low surface brightness. It lies about ⅖ of the way and a little west of the line joining α Trianguli with β Andromedae. Choose a very dark night, and allow the eye to wander generally over the field rather than to seek a single point, as is often done when chasing a nebula. A large, ghostly patch will gradually assert itself. The lower the magnification, the easier it is to see; which is why it is not a difficult naked-eye object, but is very elusive in a large telescope.

Triangulum Australe, The Southern Triangle (June 22; MAP 8)
A bright south circumpolar group, containing Milky Way fields.

Cluster N.G.C. 6025: a fine open cluster 15′ across; many stars can be resolved with binoculars.

Tucana, The Toucan (October 16; MAP 8)
A far southern group with few bright stars, southwest of Achernar, but containing the Lesser Magellanic Cloud. This is the smaller of the two satellite galaxies belonging to our system, and less spectacular than the Greater Cloud; but it is full of interesting sweeping.

Double star β: mag. 4·5, 4·5; 168°; 27″. Easily elongated with small binoculars.

Clusters N.G.C. 104, 47 Tucanae: the finest globular cluster in the sky after ω Centauri; over 20′ across, and well visible to the naked eye next to the Lesser Cloud. Appears as an extensive nebulous mass in binoculars, with a blazing center. N.G.C. 362: a smaller but still distinctive globular cluster on the edge of the Cloud.

Ursa Major, The Great Bear (or Plough) (April 11; MAPS 1, 4, 5)
The seven stars forming such a distinctive pattern in this far northern constellation mark only a part of its great area, which extends into rather barren sky both west and south of the Plough.

Double star ζ (Mizar): one of the most famous pairs in the sky, mags. 2·3, 4·0, distance 11½′, easily separated with the naked eye. Binoculars will show a mag. 8 star almost between them. The 4th magnitude star, Alcor, was held by the old Arab astronomers to be a severe test of vision, which it certainly is not today; either it has brightened, or the records are confused.

Nebulae M.81 (N.G.C. 3031) & M.82 (N.G.C. 3034): two galaxies in the same field of view. They are among the closest to us (distance about 2·6 million parsecs), and appear of very different aspect, M.81 being the brighter of the two, elongated, with a condensed nucleus. M.82, on the other hand, is very elongated and more diffuse. M.97 (N.G.C. 3587): one of the largest planetary nebulae in the sky; dim, but well seen with 12 × 40 binoculars as a circular, evenly illuminated patch with well-defined margins, very different from the centrally condensed appearance of a globular cluster or many galaxies.

Ursa Minor, The Little Bear (June 13; MAP 1)
The north circumpolar constellation, obvious from the pole star α (mag. 2·0, very slightly variable), but lacking binocular objects.

Colored star β: mag. 2·1, yellow. Note how this star appears to gain brightness in moonlight. This is caused by contrast against the bluish sky, and helped by the fact that the northern sky is relatively dark at such times.

Vela, The Sails (March 15; MAP 4)
A fine mid-southern Milky Way group lying between Puppis and Centaurus. Here are many bright clusters and good fields.

Colored stars γ: mag. 1·8, blue-white; one of the hottest stars known, with a temperature of about 18,000°C. λ: mag. 2·2, fine yellow.

Clusters N.G.C. 2547: a naked-eye cluster, partly resolved with binoculars. N.G.C. 2932: a scattered cluster, well resolved.

Virgo, The Virgin (May 10; MAPS 4, 5)
A fine, open equatorial constellation with its brighter stars well distributed, occupying the ecliptic between Leo and Libra. Its leader α (Spica, mag. 1·0) shines prominently in the south on a northern spring evening.

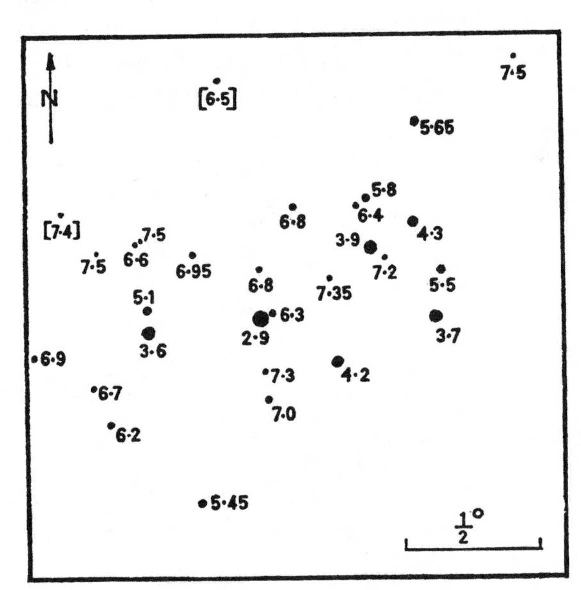

Fig. 29. *The Pleiades*

This chart shows the positions and magnitudes of all stars down to magnitude 7·5. The two stars in parentheses are not physical members of the cluster.

Double star τ: mag. 4·3, 9·6; 290°; 80″. A difficult binocular object with normal apertures.

Nebulae: The space between ϵ Virginis and β Leonis contains many galaxies; those in the bordering constellation of Coma Berenices have already been mentioned. M.49, 58, 87 and 104 (N.G.C. 4472, 4579, 4486 and 4594) should all be visible in good conditions, while M.60 (N.G.C. 4649) may also be glimpsed. An enlarged chart of these objects is given in Figure 26.

Volans, The Flying Fish (February 17; MAP 8)
A small far southern group east of the Greater Magellanic Cloud, containing no objects of particular interest.

Vulpecula, The Fox (August 25; MAPS 6, 7)
A poorly marked mid-northern group lying in the Milky Way between Sagitta and Cygnus. Fine sweeping.

Colored star α: a fine deep yellow star, mag. 4·45. About 5° south, on the border of Sagitta, binoculars will show a distinctive little "arrowhead" grouping of mag. 6 and 7 stars.

Nebula M.27 (N.G.C. 6853): the "Dumb-bell" nebula, a large planetary appearing as a misty patch about 2° southeast of star 13.

Cluster N.G.C. 6940: a very fine open cluster of faint stars, well seen in binoculars, with some 8th magnitude stars on the border.

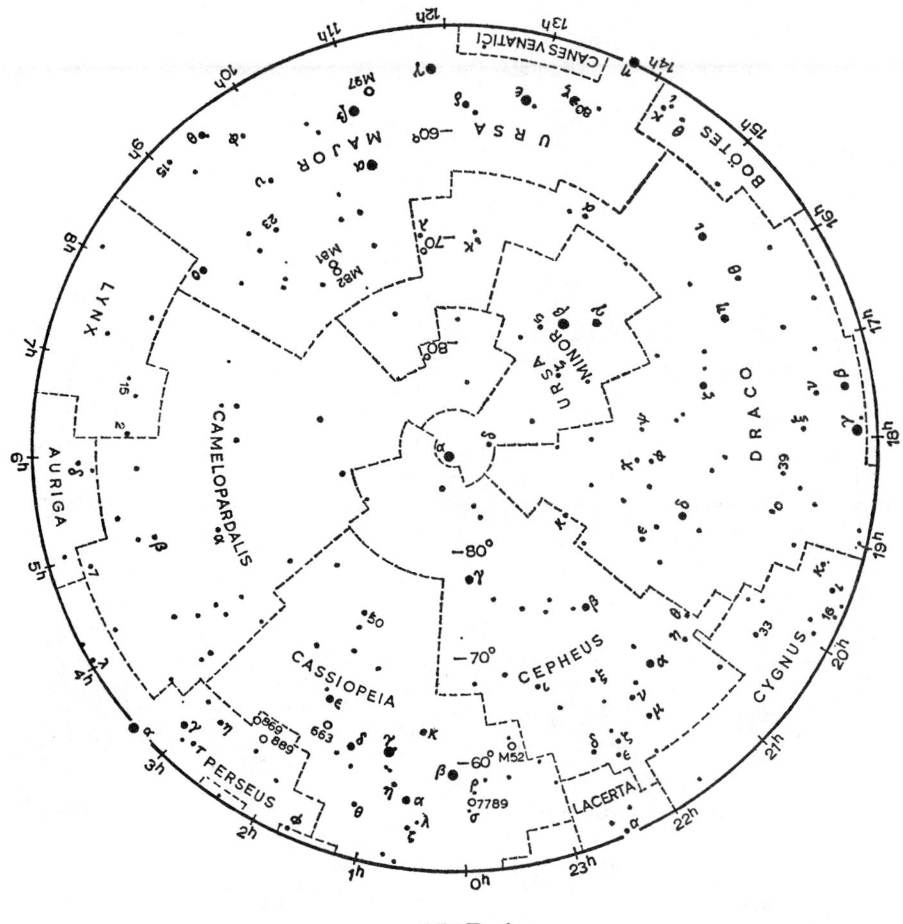
URSA MAJOR
CANES VENATICI
BOÖTES
DRACO
URSA MINOR
LYNX
CAMELOPARDALIS
AURIGA
CASSIOPEIA
CEPHEUS
CYGNUS
LACERTA
PERSEUS
M97
M81
M82
M52
663
7789
869
889
-60°
-70°
-80°

MAP 1

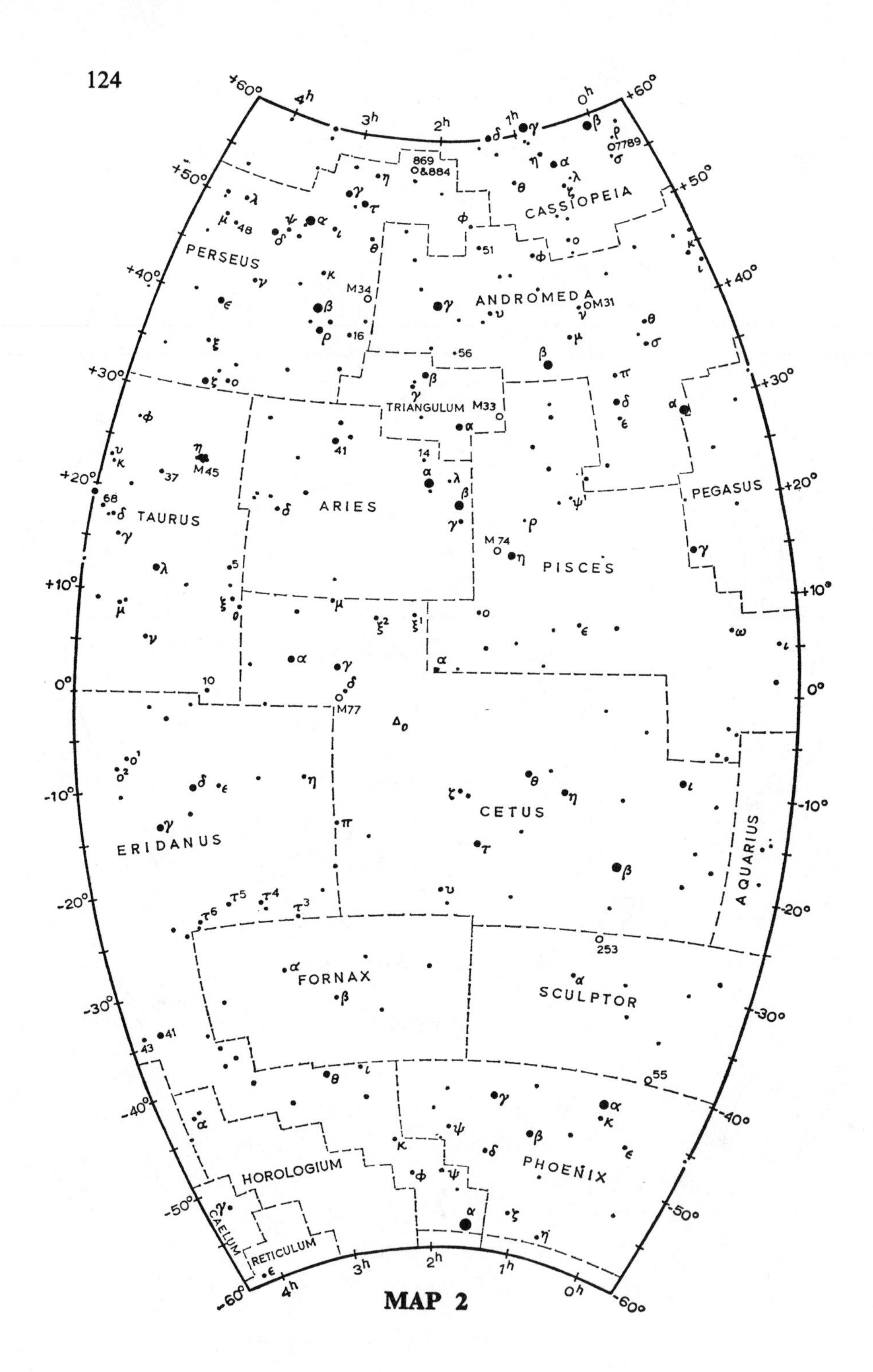
CASSIOPEIA
PERSEUS
ANDROMEDA
TRIANGULUM
ARIES
TAURUS
PISCES
PEGASUS
CETUS
ERIDANUS
AQUARIUS
FORNAX
SCULPTOR
PHOENIX
HOROLOGIUM
CAELUM
RETICULUM
M31
M33
M34
M45
M74
M77

MAP 2

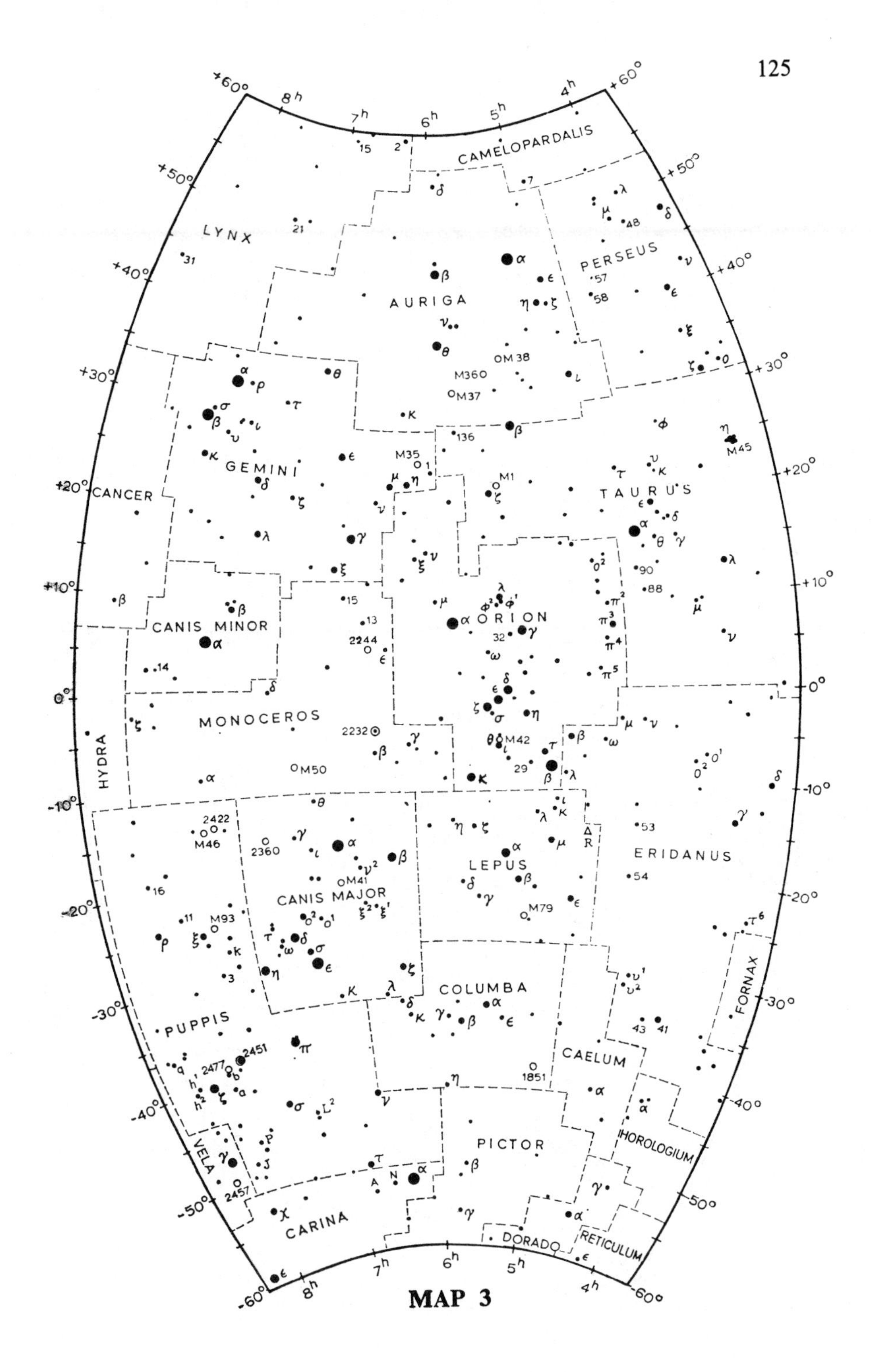
CAMELOPARDALIS
LYNX
PERSEUS
AURIGA
GEMINI
CANCER
TAURUS
ORION
CANIS MINOR
MONOCEROS
HYDRA
ERIDANUS
LEPUS
CANIS MAJOR
PUPPIS
COLUMBA
CAELUM
HOROLOGIUM
FORNAX
PICTOR
VELA
CARINA
DORADO
RETICULUM
M45
M1
M35
M36
M37
M38
M42
M50
M46
M41
M93
M79
2422
2360
2244
2232
2451
2477
2457
1851

MAP 3

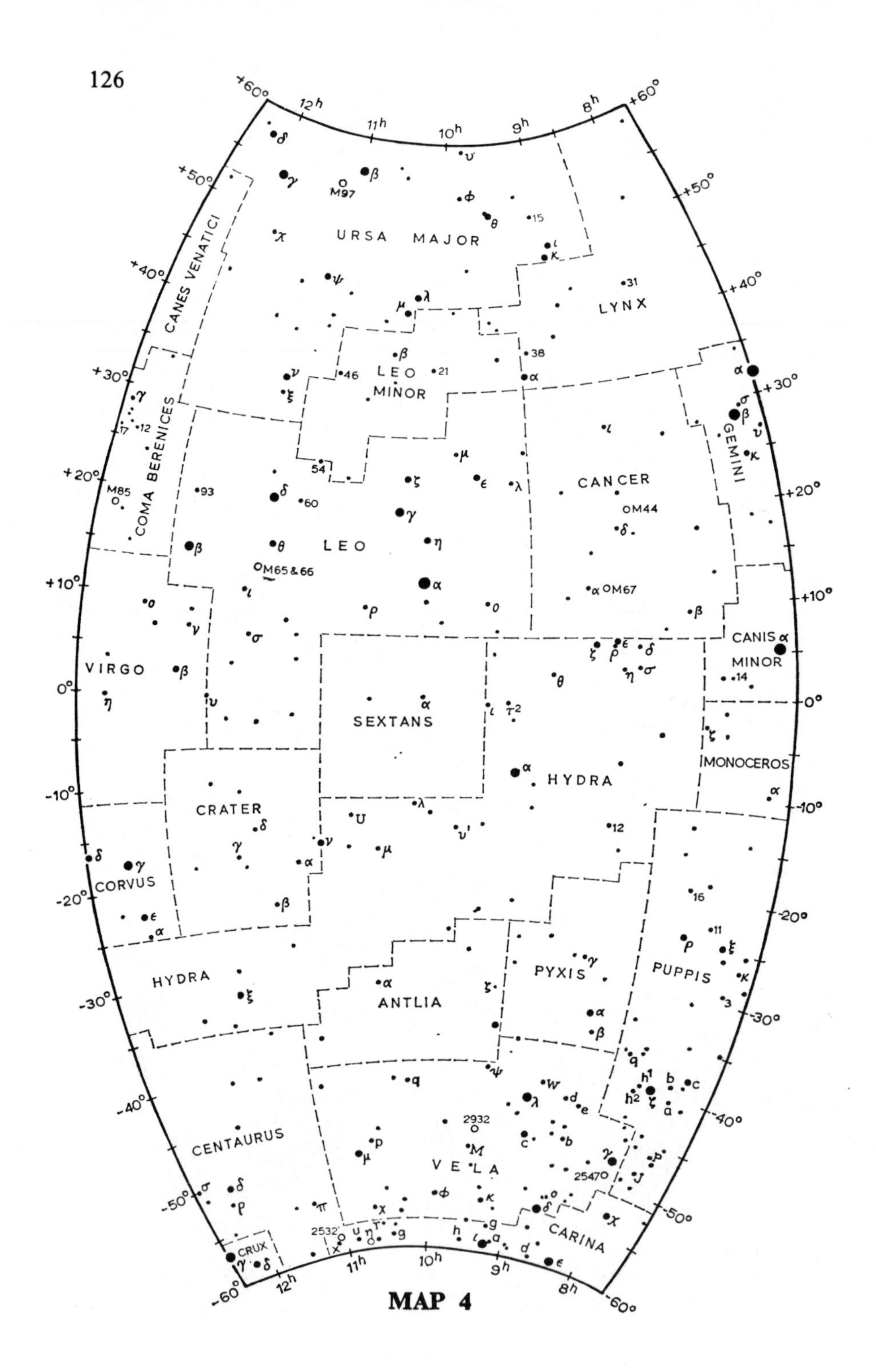
URSA MAJOR
CANES VENATICI
LYNX
LEO MINOR
COMA BERENICES
GEMINI
CANCER
LEO
VIRGO
CANIS MINOR
SEXTANS
HYDRA
MONOCEROS
CRATER
CORVUS
PUPPIS
PYXIS
ANTLIA
CENTAURUS
VELA
CARINA
CRUX
M97
M85
M65 & 66
M44
M67
2932
2547
2532
12h
11h
10h
9h
8h
+60°
+50°
+40°
+30°
+20°
+10°
0°
-10°
-20°
-30°
-40°
-50°
-60°

MAP 4

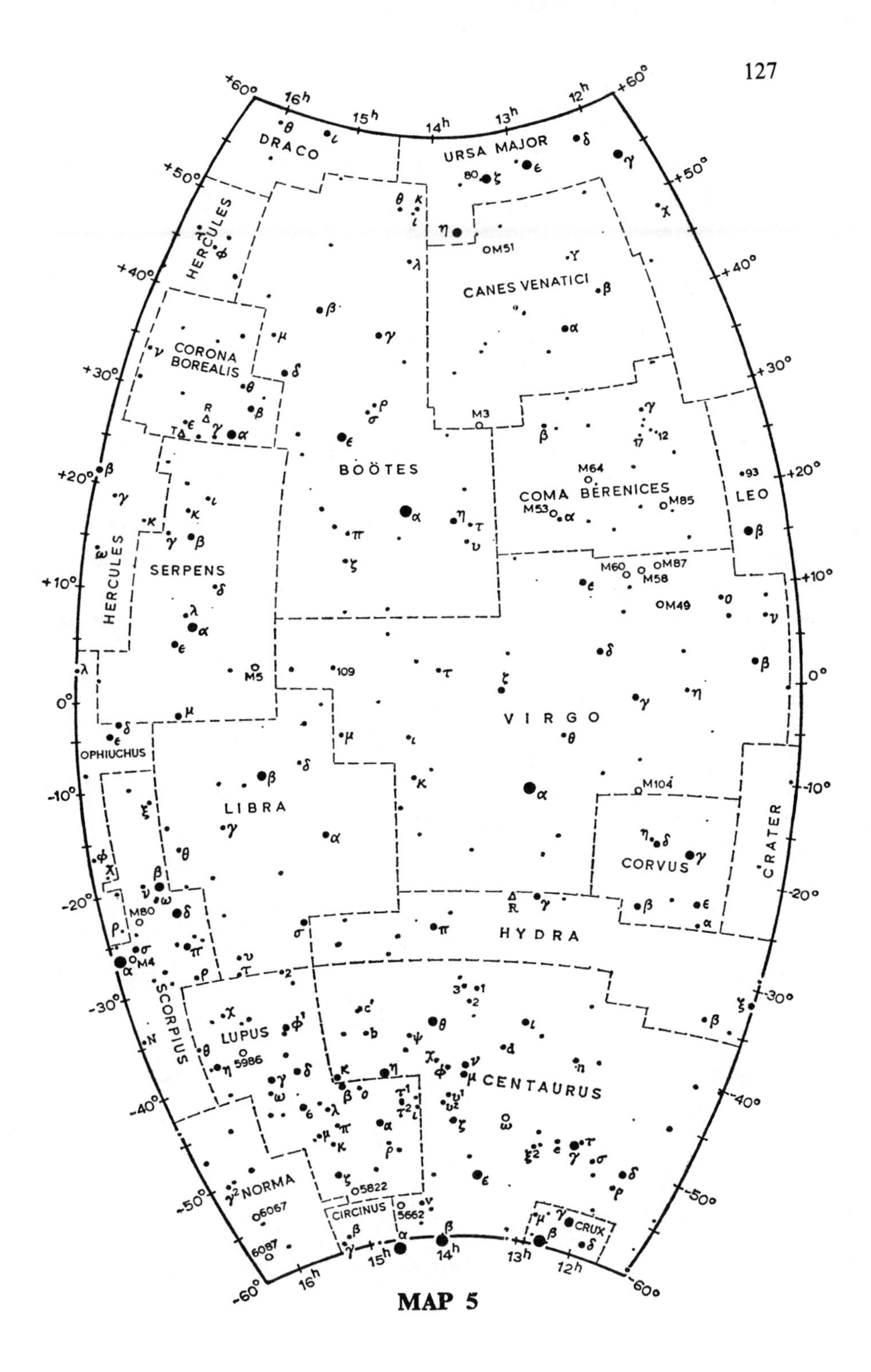

DRACO
URSA MAJOR
HERCULES
CANES VENATICI
CORONA BOREALIS
BOÖTES
COMA BERENICES
LEO
SERPENS
VIRGO
OPHIUCHUS
LIBRA
CRATER
CORVUS
HYDRA
SCORPIUS
LUPUS
CENTAURUS
NORMA
CIRCINUS
CRUX
M3
M5
M4
M80
M64
M53
M85
M60
M87
M58
M49
M104
M51
5986
5822
5662
6067
6087

MAP 5

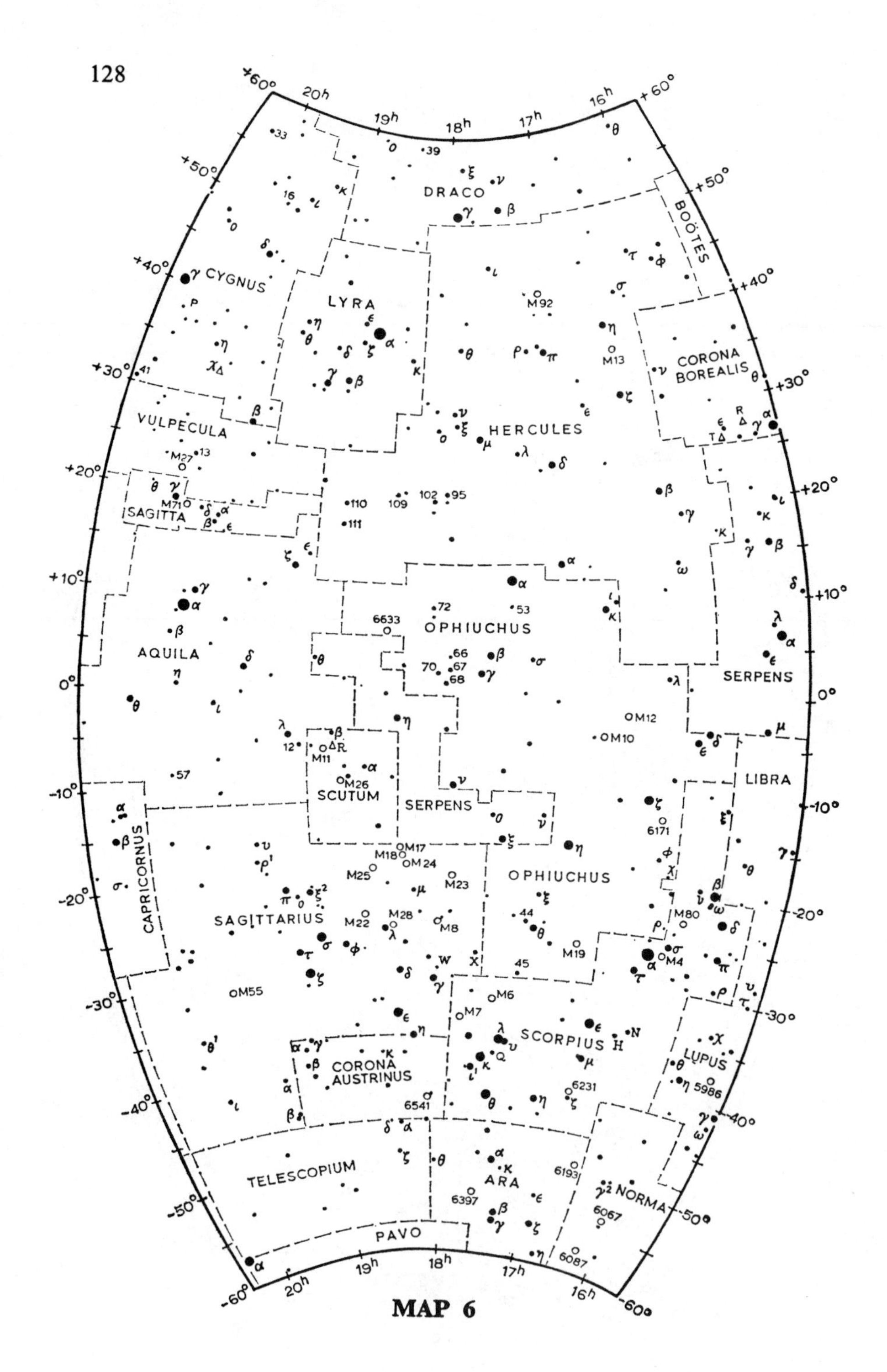

DRACO
CYGNUS
LYRA
BOÖTES
CORONA BOREALIS
HERCULES
VULPECULA
SAGITTA
AQUILA
OPHIUCHUS
SERPENS
SCUTUM
LIBRA
CAPRICORNUS
SAGITTARIUS
SCORPIUS
CORONA AUSTRINUS
LUPUS
TELESCOPIUM
ARA
NORMA
PAVO
M92
M13
M27
M71
M11
M26
M12
M10
M17
M18
M24
M25
M23
M22
M28
M8
M19
M80
M4
M55
M6
M7
6633
6171
6541
6231
5986
6193
6397
6067
6087
20h
19h
18h
17h
16h
+60°
+50°
+40°
+30°
+20°
+10°
0°
-10°
-20°
-30°
-40°
-50°
-60°

MAP 6

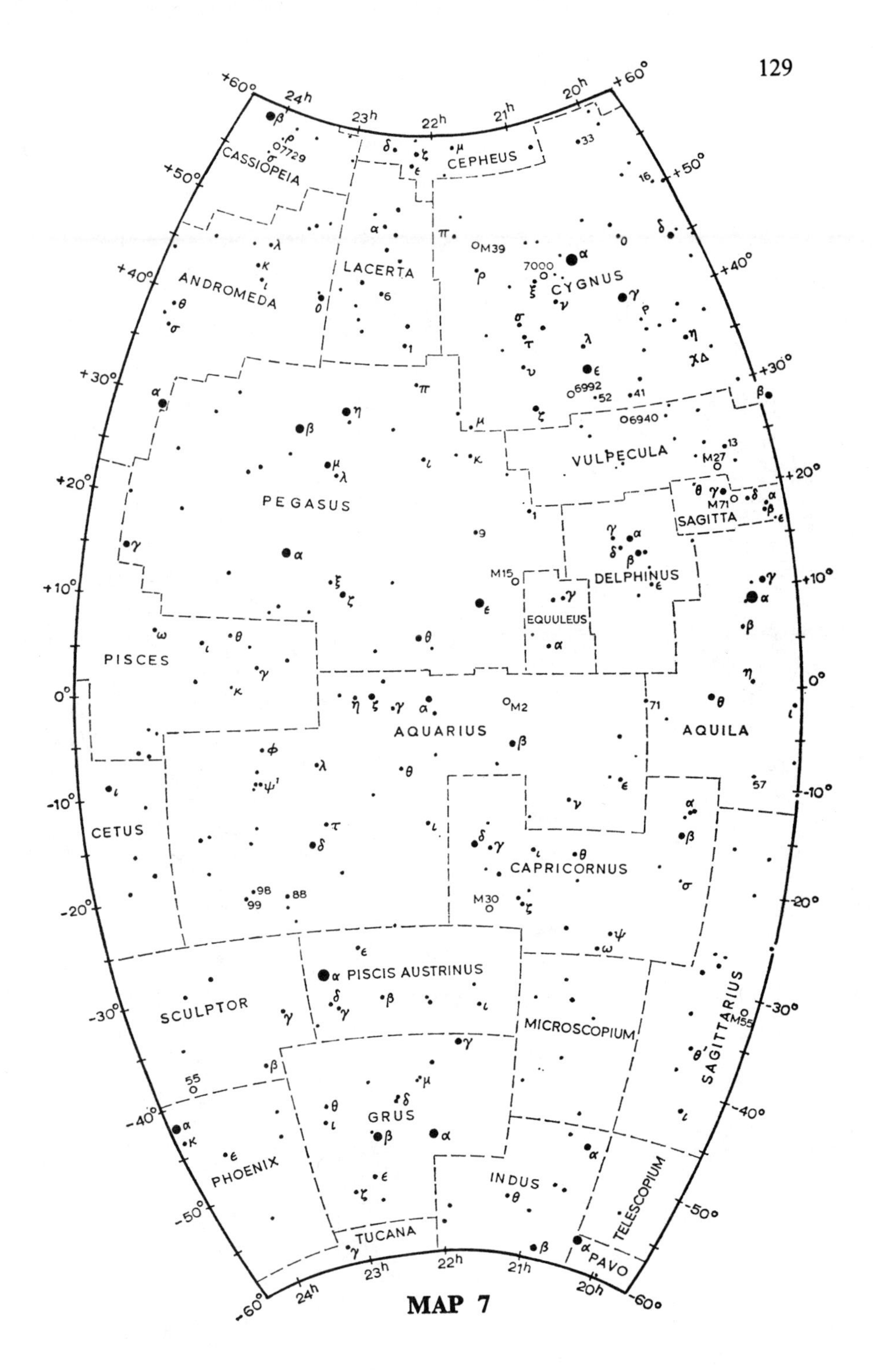
CASSIOPEIA
CEPHEUS
ANDROMEDA
LACERTA
CYGNUS
PEGASUS
VULPECULA
SAGITTA
DELPHINUS
EQUULEUS
PISCES
AQUARIUS
AQUILA
CETUS
CAPRICORNUS
PISCIS AUSTRINUS
SCULPTOR
MICROSCOPIUM
SAGITTARIUS
GRUS
PHOENIX
INDUS
TELESCOPIUM
TUCANA
PAVO

MAP 7

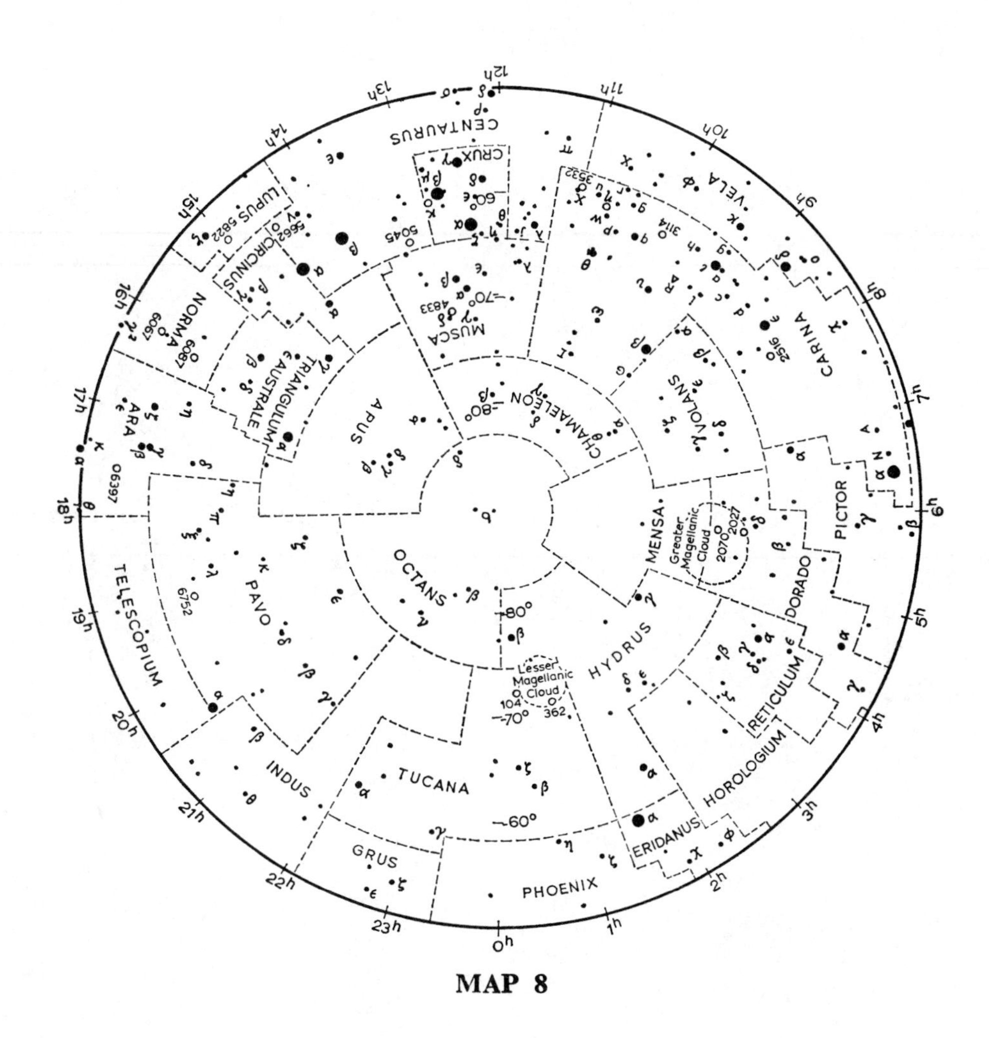
CENTAURUS
CRUX
LUPUS
CIRCINUS
NORMA
TRIANGULUM AUSTRALE
ARA
APUS
MUSCA
CHAMAELEON
VELA
CARINA
VOLANS
PICTOR
MENSA
Greater Magellanic Cloud
DORADO
OCTANS
HYDRUS
Lesser Magellanic Cloud
RETICULUM
HOROLOGIUM
PAVO
TELESCOPIUM
INDUS
TUCANA
GRUS
PHOENIX
ERIDANUS
0h
1h
2h
3h
4h
5h
6h
7h
8h
9h
10h
11h
12h
13h
14h
15h
16h
17h
18h
19h
20h
21h
22h
23h
-60°
-70°
-80°

MAP 8

☆ 8 ☆

Programs for Binocular Observers

The preceding chapters of this book have been intended to enlighten the newcomer to astronomy who wishes to take his first observational steps using a pair of binoculars or a very small telescope. I have, however, tried always to emphasize the real usefulness of modest hand instruments, and the present chapter discusses in more detail some serious programs of work for the more experienced observer who, through necessity or inclination, lacks a true "astronomical" telescope. The potential value of binoculars is evidenced by, for example, the variable-star and nova-search programs of the AAVSO, while British amateurs, using binoculars, have discovered five novae during the past decade. Giant war-surplus binoculars have also claimed many comet discoveries.

Binocular variable stars

The term "binocular variables" applies most aptly to a certain class of star, of the red semiregular or irregular type, whose light variations lie between about the 5th and 9th magnitudes and so can be followed entirely with small apertures. It is also true that binoculars are useful for observations of the bright phases of long-period variables, whose periods are fairly predictable; and for the normal phase of a couple of R Coronae Borealis type stars—but in almost all cases the range of these stars takes them beyond the grasp of most binoculars when at minimum, so that a complete series of observations cannot be obtained.

Red semiregular stars, which, like the long-period variables, are red giants many times the diameter of the Sun, are distinguished by less predictable periods and also by smaller ranges of brightness, the variation usually being between one and two magnitudes. Because of their less spectacular variation, these semiregular stars have been seriously neglected by both amateur and professional observers; and, until the recent interest in binocular observation made observers look for objects to study, hardly any were being observed on an organized basis. Now, however, binocular observers of these stars are finding that there is a real demand for data.

The likely error of a single estimate by an observer being of the order of 0·2 mag.—in adverse circumstances it may be twice this —makes it clear that scattered observations are unlikely to be of much help in following the behavior of a star whose range is only about a magnitude. By combining the results of a number of observers, however, the accuracy of the resultant light curve can be greatly improved, and fine detail of only 0·1 mag. may be distinguishable—considerably less than the probable error of a single observation. The necessity for coordinating observation of binocular variable stars—particularly in ensuring that standard comparison stars are used throughout—increases as the stars' ranges decrease, and is particularly so in the case of red irregular variables. In these stars we find that their fluctuations do not suggest even an approximate period, and the ranges are often not more than half a magnitude or so. Such stars have been little studied because of the difficulty of obtaining useful light curves; and a group of binocular observers can, once again, supply important data.

The variations of most of these stars are so slow that one or two observations a week (if the weather permits!) are adequate. Indeed, more frequent observation may lower the accuracy of the work by introducing bias—the fault in which the observer remembers his previous observations of the star and subconsciously predicts what the next observation will be. If the observer wants to see a star change noticeably in brightness during a single night's work, he should try his hand at observing an eclipsing binary with a period of about a day. The object of this work is to establish the time of minimum brightness, by making observations at 15- or 30-minute intervals around the predicted time of minimum brightness. The magnitude observations can then be plotted on a graph, and the moment of maximum eclipse established. Although the periods of such stars have usually been well determined at their time of discovery, most are then left to their own devices; and if the period shows a change, perhaps caused by some tidal interaction between the stars, it may well go unnoticed, putting the predictions many hours out as the years pass. The observer should not, therefore, be surprised to find his binary remaining obstinately at maximum when the predictions indicate "minimum!" The calculation of the times of minimum, and the observations themselves and the subsequent analysis, form a fine field of work for the inquiring observer who enjoys working independently; and it is one that has been much neglected.

To many potential observers, variable-star study sounds academic and unexciting, lacking the classical romance of lunar and planetary observation—a statistics-ridden subject where the observer becomes little more than a cipher. But then, a variable star cannot be expected to fascinate until it shows variation! Once an observer has persevered long enough to observe a change of brightness in one of these objects, he may well be converted; and if not, he has at least tried. Furthermore, there is a serious shortage of observers. Perhaps a quarter of a million variable-star observations are made annually by the world's amateurs, but the majority of them are concentrated on relatively few telescopic stars; semiregular, irregular and eclipsing stars are not, in general, among them.

Charts for several hundred of these objects, many of which have been badly neglected, have been produced by amateur groups such as the AAVSO. The availability of standard charts is important. There may be a number of stars in the field of the variable that seem suitable for use as comparison objects; but, extraordinary as it may seem, there is no comprehensive and unified catalogue of accurate star magnitudes below the naked-eye limit! Observers must perforce construct their own "sequences" of comparison stars by checking against nearby standard stars—a slow and laborious process, and one, clearly, that can lead to confusion if each observer produces his own sequence and then tries to combine his results with those of others. The distribution of standard sequences is the essential beginning to the effective study of these stars.

To illustrate this chapter, and to give the prospective binocular observer an immediate start, four interesting binocular variable stars have been selected. Each one shows a different type of variation, and has a relatively generous range of brightness.

RZ Cassiopeiae (Fig. 30) Mag. range 6·4–7·9 (white). This is an Algol (dark-eclipsing) binary, with a catalogue period of 1·1952472 days. It should be noted that periods of such precision refer to a "Sun-centered" (heliocentric) observer, and that predictions must allow for the Earth's varying distance from the star as it revolves around the Sun, which can amount to several minutes of light-time; this effect is, of course, taken into account when predictions are drawn up. John E. Isles, Director of the B.A.A. Variable Star Section, who has provided notes on these stars, comments that this is the easiest eclipsing binary for which it is possible to get a good timing of minimum from visual estimates; it is much easier than Algol.

CH Cygni (Fig. 31) Mag. range 6·0–7·8, period 97 days (red). This is one of the most interesting of all the binocular variables, and the catalogue period is open to doubt. Binocular Sky Society

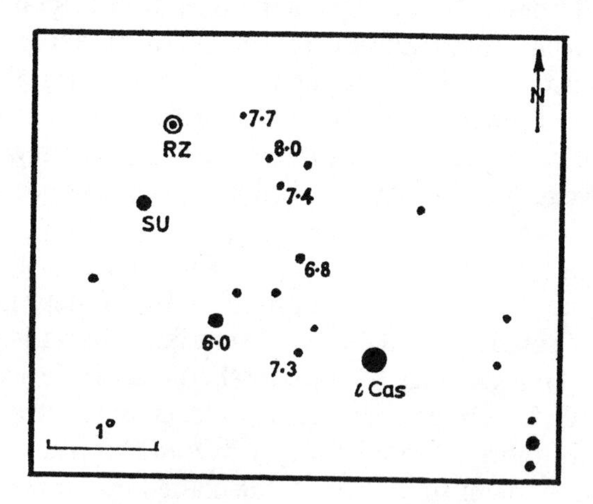

Fig. 30. *Chart for RZ Cassiopeiae*

Comparison star magnitudes are indicated. Note SU, a Cepheid variable (range 5·8–6·2, period 1·95 days).

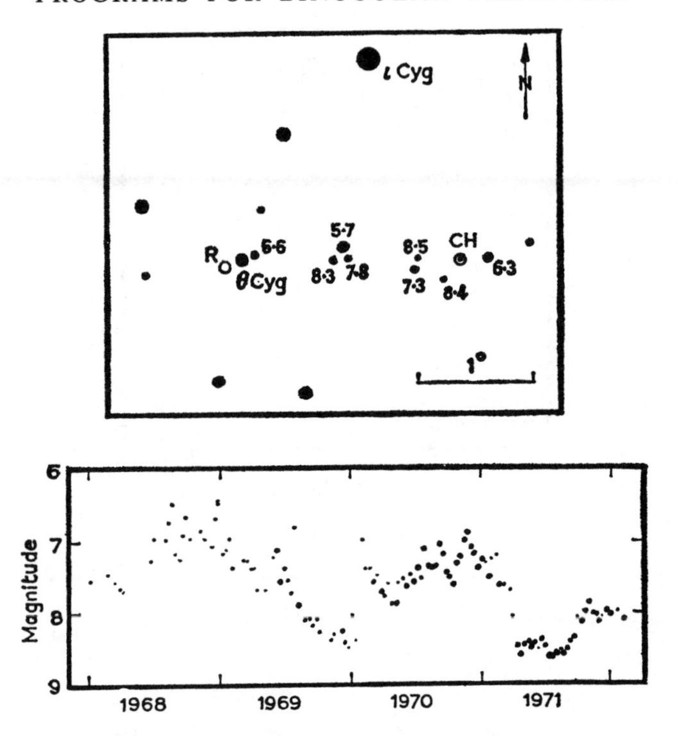

Fig. 31. *Chart and light curve for CH Cygni*

Comparison star magnitudes are indicated. In the light curve, each dot represents the mean magnitude obtained by averaging observations over 10 days; increasing size indicates 1–2, 3–4, and 5 or more estimates during this time interval. The diagram is based on one by John E. Isles from Binocular Sky Society observations.

observations for 1968–1972 indicated a rough period of 600 days, with sharp variations superimposed. There also seems to be a long, shallow brightness cycle of 4,700 days. This star belongs to a class known as the Z Andromedae stars, probably consisting of a red giant star with a small, very hot companion. Its nature is not, however, fully understood.

TX Draconis (Fig. 32) Mag. range 6·0–8·2, period 78 days (red). A semiregular variable, which in recent years has varied over practically its full range in each cycle.

R Scuti (Fig. 33) Mag. range 4·8–6·5, period 140 days (yellow). This star belongs to the RV Tauri class, and is a semiregular variable characterized by deep and shallow minima, the maxima

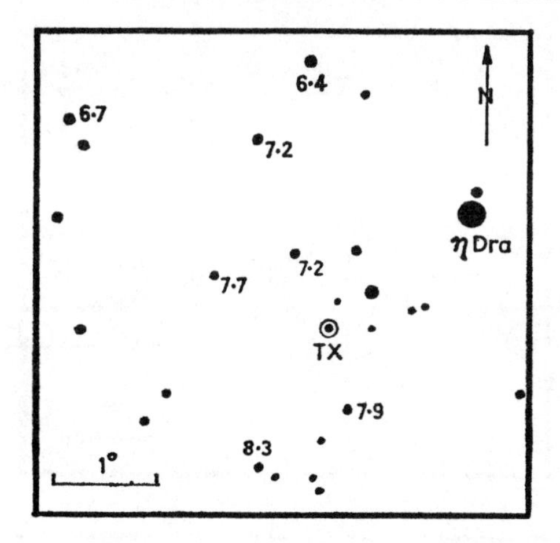

Fig. 32. *Chart for TX Draconis*

Comparison star magnitudes are indicated.

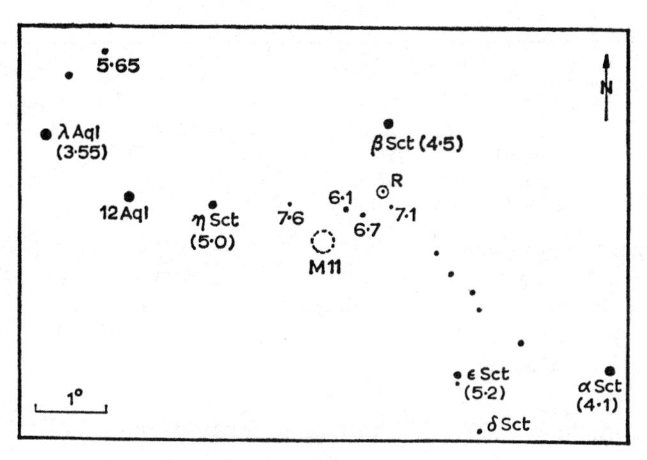

Fig. 33. *Chart for R Scuti*

Comparison star magnitudes are indicated.

being more constant in brightness. The period refers to a complete cycle of two maxima and minima. The RV Tauri stars are believed to be very luminous giant stars going through a short-lived and complex instability, giving rise to this double pulsation.

When making variable-star estimates, a number of points must be borne in mind. Basically, the eye is a poor instrument for this work. Place two equally bright stars in the field of view and it may judge equally of each; but it is more likely that one will be considered the brighter, possibly by several tenths of a magnitude. Move the stars in relation to each other, and the other may now appear to be the brighter! Change the color of one star, while keeping the light intensity the same, and another interpretation will follow.

There is space here only to outline some simple rules, which, if followed, will at least curtail the eye's lack of discipline.

1. Bring the variable and the comparison star to the center of the field in turn, take one quick look (not more than a couple of seconds), and memorize the relative brightnesses.

Never look "in between" the stars to be compared, to try to judge each simultaneously. The sensitivity of the retina varies from place to place; by bringing each star to the center of the retina, consistent sensitivity is helped.

Never stare at a star for a long period. The eye's judgment of brightness will not be improved, and, if the star has a reddish tint, it will appear to grow brighter with prolonged staring.

Bringing each star to the center of the field also minimizes the effect of marginal blurring in the field of view, and to a large extent vitiates the "position angle error" mentioned above, where the stars' relative apparent brightness is affected by their positions in the field of view.

2. Decide first of all which star appears the brighter. Once this has been done, the observer stands on solid ground on which to refine his estimate.

3. Aim at bracketing the variable in between two comparison stars. Then define the variable's brightness either by taking steps of 0·1 mag. from each comparison star, or by making a fractional estimate of its brightness in the total magnitude interval separating the comparison stars.

Suppose that the comparison stars bracketing the variable are of mag. 6·2 and 6·7. The observer might judge the variable to be 0·2 mag. fainter than one, and 0·3 mag. brighter than the other. The estimate would be written as

$$6{\cdot}2 - 2,\ 6{\cdot}7 + 3 = 6{\cdot}4.$$

Note that variable-star arithmetic differs somewhat from that taught in schools, the reason being that the *lower* magnitude number refers to the *brighter* object.

The observer using the fractional method would tackle the problem in a different mental way, by dividing the brightness difference between the mag. 6·2 and 6·7 stars into a number of intervals, so that the variable fell into one of the intervals. He might judge the variable to be one third of the way down from the 6·2 to the 6·7 star. The estimate would then be written

$$6{\cdot}2_1V_26{\cdot}7 = 6{\cdot}37,$$

which would, of course, be rounded off to 6·4. It is conventional, when writing down variable-star estimates, to put the brighter comparison first.

In general, the fractional method is useful where intervals of more than about half a magnitude are involved, at which point it becomes difficult to evaluate the number of 0·1 mag. steps in the sequence.

4. If the color of the variable is red, which will normally be the case, some of the natural difficulty of comparing it with a white comparison star can be eased by putting the binoculars slightly out of focus, so that the stars are expanded into disks. The diminution of intensity may be sufficient to bring the light level below the threshold of the eye's color sensitivity. It will, at all events, reduce the intensity of the color. Some observers defocus as a general rule when the brightness of the stars will allow them still to be seen, finding it easier to compare the relative intensities of disks than of points.

Nova hunting

Every year, perhaps several dozen of the more massive stars in our galaxy pass through an explosively unstable phase. When this happens, the outer shell of the star blasts outwards and the brightness rises by up to 12 magnitudes, usually in only a few hours. These novae cannot be predicted, since the stars in their normal state are usually extremely faint objects.

The nova phenomenon is both spectacular and important, for astrophysicists can here learn much about the behavior of matter under extreme conditions, as well as understanding better the place of the novae in the stellar life cycle. Observations of the rise to maximum are, because of the brevity of this phase, particularly scarce. Many novae are, however, discovered after their outburst by professional astronomers scanning old photographic plates (very possibly with some completely different aim in mind!), and little useful information may be obtained. What is wanted are "live" discoveries, made on the very night of maximum, so that observations with powerful telescopes can be made at once.

Here, then, is a fine field of observation for the patient amateur who is prepared to scan the sky on every clear night. One advantage recommending the work to many amateurs is that scattered cloud, haze, or moonlight can still permit a search for bright novae to take place, so that the most can be extracted from our generally poor conditions. The Milky Way region is the most profitable area in which to look, for it is here that we are seeing the greatest number of stars per unit area, and relatively few novae have been observed away from its general limits. Some novae reach naked-eye brightness, and require no instrumental aid at all for their discovery—during this century, out of 158 novae observed in our galaxy (a very small percentage of the total occurring), 16 were brighter than mag. $4\frac{1}{2}$ at maximum, and nearly 60 would have been easily observed in a pair of 8×30 binoculars.

Clearly, no nova discovery can be made without an excellent knowledge of the sky; but this is a knowledge that all amateurs, regardless of their particular interest, should aim at acquiring. A sound knowledge of the constellations down to mag. 4, and a regular naked-eye survey, is all that is needed to catch a bright nova. An excellent way of learning the sky to this level of intimacy is to take a couple of constellations every night and to learn their lettered stars at the rate of six at a time (α to ζ on the first night, η to μ on the next, and so on). Routine sweeps will not need to be carried consciously to every star, for the unusual appearance of a familiar pattern will register at once on the observer's brain.

Surprisingly few observers carry out even this simple patrol, although it takes only ten minutes or so and could earn them immortality, besides being a pleasant relaxation. But a few amateurs have gone so far as to make binocular nova-sweeping their main work, spending an hour or two every evening scanning the Milky Way down to the 7th or 8th magnitude. The leading amateur nova hunter is the British observer G. E. D. Alcock, who made binocular discoveries in 1967, 1968, 1970 and 1975. He commenced regular sweeping in 1953, and holds the world record for visual nova discoveries.

Once one goes to the trouble of sweeping for novae with binoculars, the question naturally arises of whether the work can be reduced by selecting "nova-prone" regions and ignoring the less productive parts of the Milky Way. An examination of the catalogues will reveal many novae in the constellations of Aquila, Sagittarius and Scorpius, and few in Auriga, Gemini and Perseus, although they are of roughly the same area. The reason is that the former groups lie in the direction of the galactic center, so that we see a greater depth of stars than in any other part of the Milky Way. On the other hand, many of the novae in this direction are too distant and faint to be discoverable with binoculars in any case. Figure 34 indicates the way that the fainter novae appear to cluster around the galactic center; but it should not be overlooked that this region of the sky has come under particularly close examination, and there may be many equally faint novae in other directions that were not picked up.

General rules for nova hunters would seem to be the following:

1. Bright novae (mag. $4\frac{1}{2}$ and above) show a general distribution around the Milky Way, which should, therefore, be given a nightly naked-eye search.

2. Most novae brighter than mag. $7\frac{1}{2}$, which is about the practical limit for checking with 8×30 binoculars, have been detected in that part of the Milky Way lying between Cepheus in the north and Crux in the south, with Sagittarius-Scorpius at the center. Northern observers would do well to afford priority to the Milky Way from Sagitta to the southern horizon.

3. Observers with a good view of the galactic center (south of about latitude 35°N), could undertake a concentrated search for novae down to the 9th magnitude, using, for example, 10×50 binoculars. Taking γ Sagittarii as the center, a strip 10° wide extending about 20° along the Milky Way will probably produce at least a couple of novae per annum above this magnitude limit. As far as the writer is aware, no such work is currently being undertaken; but it could be done by a patient observer with every hope of success. Some amateurs in New Zealand are currently scrutinizing the Magellanic Clouds for novae.

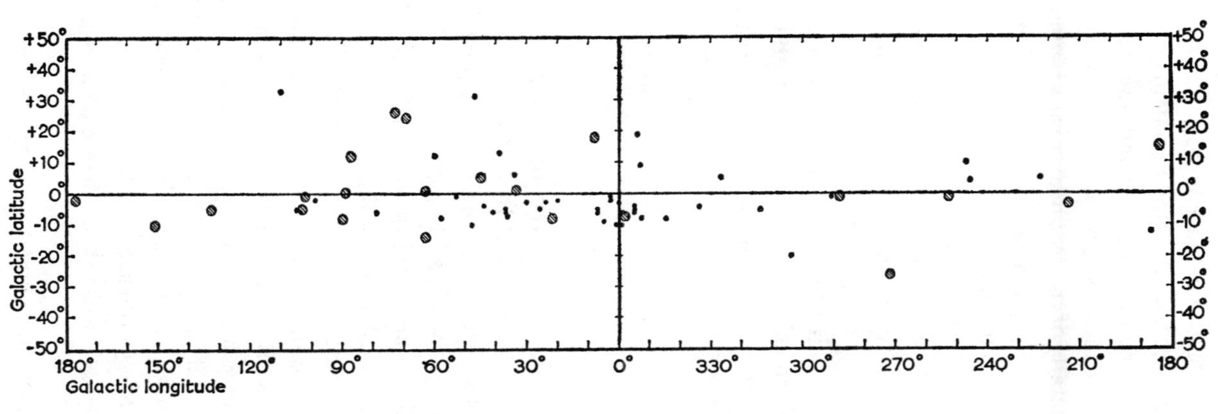

Fig. 34. *The distribution of galactic novae*

Dots indicate novae reaching magnitude $4\frac{1}{2}$–$7\frac{1}{2}$ at maximum; shaded circles indicate novae brighter than magnitude $4\frac{1}{2}$. The vertical scale represents galactic latitude, or angular distance away from the median line of the Milky Way. The horizontal scale represents galactic longitude, the center (0°) lying in the constellation Sagittarius. Most novae have been observed between longitudes 350° and 100°—from Scorpius to Cepheus.

The observation of a suspicious object which is not marked on the reference atlas is not necessarily the signal of success. The major and minor planets are constantly changing position; the regular observer will not need to be told where the major planets are to be found, but for details of the brighter minor planets, which may stray into a nova hunter's fields, the *Observer's Handbook* or some other comprehensive almanac should be consulted. On top of this, no star atlas is complete, and there are always at least a few omissions of stars that ought to be included. The observer should therefore attempt to reinforce his atlases with actual sky photographs, whose accuracy cannot be doubted. If a star is seen in the sky which is not on a photograph, and if it cannot be reconciled with the position of a planet or of a variable star, the fact should be communicated at once to the appropriate authority: normally, the national observatory.

Telescopic meteors

The nova hunter or variable-star observer, or in fact anyone who observes regularly with binoculars, will from time to time notice a meteor flash through the field. Such meteors are usually faint, of between 5th and 8th magnitudes, and so could not be seen with the naked eye. These telescopic meteors form a fruitful field for study, and one that is still largely unexplored.

As might be expected, fainter meteors generally outnumber their brighter brethren in the same way as the fainter stars exceed the naked-eye ones. But the distribution of bright and faint meteors in the well-known showers is not regular. The Quadrantid shower is active telescopically, while the Perseids are not, and even during the time of high shower activity telescopic rates are much lower than the naked-eye ones. Sporadic rates, however, are about the same as naked-eye rates: about 6–8 meteors per hour, under good conditions.

It might, at first sight, be expected that telescopic rates would depend upon the aperture of the telescope used, but this is not the case, for the reduced fields of view available with larger instruments cut down the amount of sky that can be surveyed. However, the increased faintness of the meteors visible compensates for this, and leaves the rate generally unaffected.

There is plenty of scope for binocular work on faint meteors, both in the documentation of sporadic meteors and in the detection of new showers, which may vary from year to year due to clumping effects in the meteor stream. A pair of 10 × 50 binoculars are probably best for this work, provided the field of view is between 5° and 6°. The field to be surveyed is selected in advance —an altitude of between 45° and 60° will probably prove most convenient—and a tracing is made from a suitable atlas. The ideal field will have a convenient range of stars to act as magnitude indicators for any meteors seen, and they will be well scattered, to act as reference points for reproducing the meteor's path.

The observer, once seated in his chair with the tracing on his lap and an observing pad, pencil and red lamp to hand, is ready to start observing a class of meteors inaccessible to radar or the fast professional cameras that had, until recently, partly displaced the visual observer. His task need no longer be confined to the recording of magnitudes and hourly rates. His concentration on a small and familiar field, and the magnification of the binoculars, enables him to gather other useful data with significant accuracy. These include:

1. *Path.* Make an instant mental note of the meteor's flight relative to at least two stars, preferably three. The occurrence of a train, even a momentary one, will help considerably.

2. *Duration.* The note "slow/medium/fast" is useful, but some observers achieve more quantitative results by using a stopwatch to reproduce the remembered duration, taking the average of several attempts. Naturally, telescopic meteors pass quickly through the field of view, and a duration of more than a second is rare.

3. *Brightness.* This is compared with that of field stars, whose approximate magnitudes can be obtained by comparison with some nearby variable-star sequence. It is, of course, possible to choose the field of a binocular variable for this work. The light is unlikely to be constant throughout the meteor's traverse, and a rough light curve can be drawn to give some idea of the manner of its variation.

4. *Path length.* Meteors do not often both appear and burn out inside a binocular field; they either fly right across, or else cross a border. In these cases the path length cannot be determined directly, although a large number of observations can be analyzed statistically to produce results. A common coding for the different possible circumstances is:

OO entered and left field
Oa entered and ended inside field
aO began inside field and left field
aa began and ended inside field (give length in degrees)

5. *Trains.* These are frequently seen in the case of the brighter meteors, and should be noted, although they are rarely more than momentary.

Every binocular observer should form the habit of noting down details of any telescopic meteor that may be seen, no matter what work was being conducted at the time, and the results reported annually to the appropriate center. If everyone did this, a notable increase in the available data on these objects would result! All that need be noted is the position of the field (the nearest known star can be used, the reduction to right ascension and declination being made later); the diameter of the field of view; the limiting magnitude at the time; the path details; the position angle of the direction of motion, referred to the north point of the field; the approximate magnitude of the meteor, together with notes on its velocity and color (if any); and the instrument used.

In general, the main aim of telescopic meteor work should be to obtain an accurate path, so that deductions can be made concerning any possible shower activity. Sporadic meteors will be moving in random directions, but if several meteors are seen to be moving in the same direction it is possible that they are members of a stream. To check this, move to a new field about 45° away from, and at right angles to, the direction of motion. If a shower really is in operation, meteors in this second field will be showing a new preferential direction. By plotting the directions on the atlas, and backtracking, the position of the radiant can be found with considerable accuracy, and almost certainly of a higher order than that possible with the naked eye on bright showers. Reports of any such activity should be forwarded at once to a responsible body such as the American Meteor Society.

When observing known showers, fields should be chosen at from 15° to 30° from the radiant, and at right angles to it, so that alternative viewing positions produce trails that cross each other near an angle of 90°, allowing the greatest possible precision in defining the radiant.

In a field of study as new as this one, where so little has been published, the observer must perforce draw up his own textbook. A couple of hours' watching on a clear, dark night will tell him whether the work seems agreeable. If so, and if he is prepared to spend many cold (but some warm) nights staring at stationary fields, he will have the satisfaction of knowing that he is tapping a good deal of new information. New, perhaps transient, showers can occur at any time, and might pass unknown but for the patient vigils of some amateur with binoculars or a small telescope.

Appendix 1: Astronomical Societies

The newcomer to astronomy will learn much from books and even more from outdoor observing sessions, but another essential part of one's education comes from meeting other amateurs, asking questions and engaging in cooperative work.

Anyone fortunate enough to have an active local astronomical society near at hand should join it straightaway. Then the tyro will be able to tap the experience of established observers, who, remembering their initial struggles, will be more than pleased to pass on advice. A good society will encourage its members by setting up mutual observing programs, so that the results can be compared at the next meeting; the discussions following work of this sort will be of enormous benefit to the newcomer who, working on his own, lacks meaningful standards at which to aim. It may have instruments available for loan to newcomers; it will certainly include people who can advise over the choice of binoculars and telescopes, and perhaps thwart an unwise purchase!

There are a great many regional societies in the United States—too many to list here—but a comprehensive listing appears in my *Amateur Astronomer's Handbook* (T. Y. Crowell, Publishers, 2nd edition, 1974). The two major national societies are the American Association of Variable Star Observers (AAVSO) at 187 Concord Avenue, Cambridge, Mass. 02138, and the Association of Lunar and Planetary Observers (ALPO), c/o The Observatory, New Mexico State University, Las Cruces, New Mexico 88001. Mention must also be made of the American

Meteor Society, a small organization but one that welcomes inquiries by *experienced* observers desiring to contribute to its work. Inquiries should be addressed to: American Meteor Society, D. D. Meisel, Director; Department of Physics & Astronomy, Geneseo, N.Y., 14454.

Appendix 2: Publications for the Binocular Observer

Star maps and catalogues

HOFFLEIT, D. *Yale Bright Star Catalogue*. New Haven, Conn.: Yale University Observatory, 1982 (4th edition). This new edition of a famous work lists magnitudes, positions, and physical data for over 9,000 stars brighter than mag. 6.5.

LAMPKIN, R. H. *Naked Eye Stars*. Edinburgh: Gall & Inglis, 1974. This is a very useful little book, giving the name, position, magnitude and spectral type of every star in each constellation down to mag. 5.5

NORTON, A. P. *A Star Atlas and Reference Handbook*. Cambridge, Mass.: Sky Publishing Corp., 1981 (16th edition). This standard publication, showing stars over the whole sky down to mag. 6, and including many lists and tables, should be one of the amateur's first purchases.

SCOVIL, C. E. *AAVSO Variable Star Atlas*. Cambridge, Mass.: Sky Publishing Corp., 1981. This is, essentially, a whole-sky atlas showing stars as faint as mag. 9½, but it identifies over 2,000 variable stars, and marks the magnitudes of convenient comparison stars. The positions of deep-sky objects are also indicated.

TIRION, W. *Sky Atlas 2000.0*. Cambridge, Mass.: Sky Publishing Corp./Cambridge University Press, 1982. A very useful large-format atlas covering the whole sky down to mag. 8.0, showing about 43,000 stars as well as 2,500 deep-sky objects. It supersedes the well-known *Atlas Coeli* (*Atlas of the Heavens*) by A. Becvar, which had been widely used by amateurs since its appearance in 1956.

VEHRENBERG, H. & BLANK, D. *Handbook of the Constellations*. Cambridge, Mass.: Sky Publishing Corp., 1973. Each two-page spread shows an area of sky with stars down to mag. 6, with a detailed listing of interesting objects.

Yearbooks and periodicals

Astronomy. A monthly magazine of amateur and professional astronomy. Published by AstroMedia Corp., 411 E. Mason Street, P.O. Box 92788, Milwaukee, WI 53202.

Handbook of the British Astronomical Association. This annual publication gives details of the movements of the Sun, Moon, planets, satellites, and comets; occultations; meteor showers, and much other information. Available to nonmembers of the Association from Burlington House, Piccadilly, London W1V ONL, England.

Observer's Handbook. This annual compendium is somewhat similar in coverage to the B.A.A. *Handbook*. It is published by the Royal Astronomical Society of Canada, and can be obtained from Sky Publishing Corp., 49-51 Bay State Road, Cambridge, MA. 02238–1290.

Sky & Telescope. This world-circulation monthly astronomical magazine can be recommended to all observers, no matter what equipment they have. It is published by Sky Publishing Corp. (see above).

Appendix 3: Sources of Books and Equipment

AstroMedia
PO Box 92788, Milwaukee, WI 53202

Astronomy Book Club
Riverside, NJ 08075

Book Faire
Sky Publications Corp., 49-62 Bay State Road, Cambridge, MA 02238-1290

Edmund Scientific Company
101 E. Gloucester Pike, Barrington, NJ 08007

Herbert A. Luft
69-11 229th Street, Box 91, Oakland Gardens, NY 11364

Optica b/c Company
4100 MacArthur Blvd, Oakland, CA 94619

Optron Systems
704-706 Charcot Avenue, San Jose, CA 95131

Orion Telescope Center
PO Box 158, Santa Cruz, CA 95061

Tomlin Industries Inc.
679 Easy Street, Simi Valley, CA 93065

William-Bell, Inc.
PO Box 3125, Richmond, VA 23235

Appendix 4: The Return of Halley's Comet

In the year 1986, the most famous comet of all returns to perihelion, passing closest to the Sun in the early hours of February 9. Nobody with even a passing interest in astronomy will need to be told that Halley's Comet is on its way; in addition, millions of people with no knowledge of astronomy at all will be anxious to view this legendary visitor as it passes through the warm environs of the inner planets.

Past returns

A noble reputation precedes the comet. Like some great pendulum, ticking away the fourscore years of a long life, it has now been seen across a span of twenty-nine human ages. It was first recorded by Chinese astronomers at the return of 239 B.C. When Halley's Comet made its famous appearance in the year of the Battle of Hastings, 1066, it had already been recorded at seventeen previous returns! Not until its twenty-sixth return, in 1682, did Edmund Halley realize that its orbit coincided very closely with the orbits of the comets observed in 1531 and 1607. Such an analysis was possible only because Newton's theory of gravitation offered a means by which the motions of the bodies in the solar system could be understood. Halley's prediction of a further return in 1758 was vindicated, the comet being rediscovered on Christmas Day of that year.

Our "modern" view of Halley's Comet, therefore, extends over only two apparitions: those of 1835 and 1910. At the first of these, astronomers had yet to measure the distance to the nearest stars; at the second, the nature of the galaxies, and the scale of the universe, was as yet unknown. There was no workable theory of stellar evolution, since understanding of atomic structure was in its infancy. The transformation of scientific knowledge embraced by the comet in its last orbital round does not need to be emphasized.

Halley's Comet and the public

Some people will not be reading this book until after Halley's Comet has passed its brief encounter with the inner planets, and is hurtling back towards the outer regions of the solar system, where its next, invisible aphelion in the twilight of Neptune's realm will occur in the year 2021. They may well think back and say: "What was all the fuss about?"

Nothing is so easy to lose as a reputation, and Halley's Comet has a reputation second to none. In ancient times, it was bright enough to survive in the chronicles of the time. At the return of 837, it seems to have been not much fainter than Venus. But at the return of 1910, when the geometry was very similar, the comet appeared much fainter. Perihelion passage makes enormous demands on a comet, and with each return to the Sun there is less of the volatile material remaining to go to forming part of the spectacular tail.

The 1986 return is not a favorable one, and the comet will be less widely observable than in 1910. Town dwellers may not be able to make out Halley's Comet at all, from their glare of lights. And here lies the second, more general problem: our vast urban communities are being denied the sights that were so familiar to our ancestors. The fault, then, is not entirely that of Halley's Comet. We have to face the fact that there may never be another Great Comet to amaze the world at all! There is a famous painting of Donati's Comet of 1858, hanging with its great tail in the sky over Paris. If Donati's Comet returned now, no Parisian would think it even worth a glance: it would be blotted out by lights.

Of course, astronomers will be blamed for not producing a better Halley's Comet. But it would be pleasant if the reputation of this famous object was too powerful to be destroyed entirely. Is it too much to hope that high authority would acknowledge this reputation with a token of respect? Will some of our great cities relent, and dim their lights, however briefly, so that the passage of another age may be marked by their populations?

The orbit of Halley's Comet

The distance of Halley's Comet from the Sun ranges from about 3,280 million miles (thirty-five times the Earth's distance) at aphelion, to only 55 million miles (about half the Earth's distance) at perihelion. In planetary terms, this takes the comet from beyond the orbit of Neptune to between the orbits of Mercury and Venus. This orbit is inclined at an angle of 18° to that of the Earth (represented in the sky by the ecliptic), and the comet moves around the Sun in a retrograde direction, from east to west. Almost all of its time is spent south of the ecliptic; it is north of the Earth's orbital plane for only some three months around the time of perihelion, out of a total orbital period of about 76 years.

The solid part of Halley's Comet is a tiny frozen mass some 3.1 miles across, much smaller than many asteroids. Since the comet has to cross the region of all the larger planets, it is easily affected by their gravitational pull, and these slight irregular perturbations create slight changes in its orbit from one passage to the next. Halley's Comet took only 74½ years to pass from perihelion in 1835 to that of 1910, but as much as 79¼ years elapsed between the perihelia of 451 and 530. The interval between the perihelia of 1910 (April 20) and 1986 (February 9) is 75 years 9½ months.

Appearance and movements

Let us imagine that we are looking down on the solar system from some great distance to the north of the Earth's orbital plane (see figure). The Earth's orbit appears practically a circle, which can be divided up into *ecliptic longitude*, with the Sun at the center of the "dial." Intervals on this circle are calibrated with reference to the Earth's own annual circuit. Longitude 0° marks the Earth's position at the autumnal equinox, around September 23; the winter solstice (around December 21) occurs at longitude 90°, and so on.

The orbit of Halley's Comet, projected on to this plane, is also shown. The solid line indicates the portion to the north of the Earth's orbital plane; the dashed line shows where it lies to the south. It can be seen that the perihelion point lies near longitude 310°. Our view of

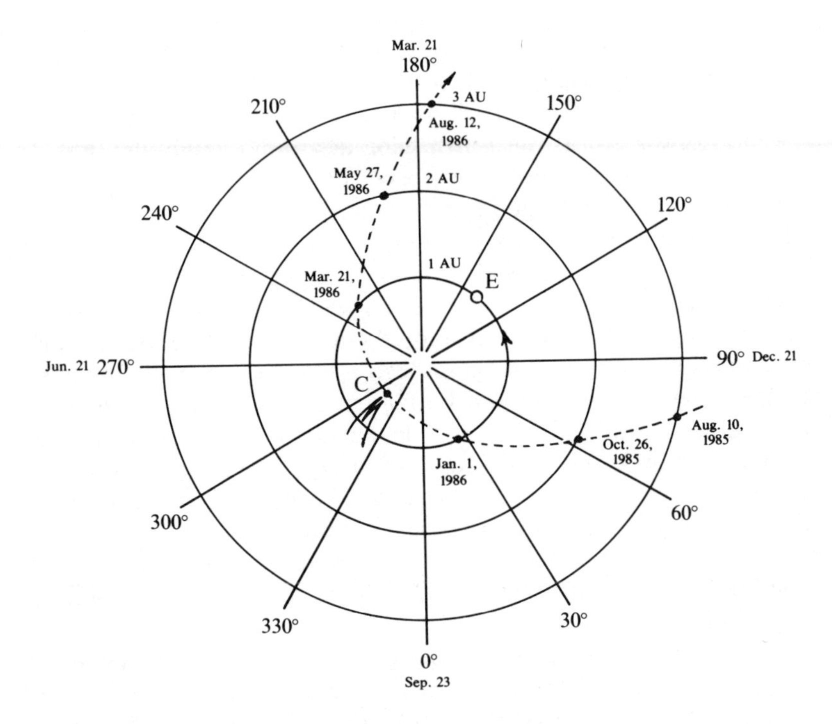

The perihelion passage of Halley's Comet. C and E indicate the position of the comet and the Earth, respectively, on the perihelion date of February 9, 1986. The dates when the comet is at the distance of 1, 2, and 3 astronomical units (AU) from the Sun, both before and after perihelion, are also shown.

the comet at perihelion obviously depends upon the position that the Earth happens to occupy in its orbit at the time.

In 1986, perihelion occurs in early February, when the Earth is at longitude 140°. This is the most unfavorable possible time, since the comet will be on the opposite side of the Sun, and therefore totally invisible. It will disappear into the Sun's rays, in the evening sky, in the second half of January, and reappear in the morning sky about a month later. If the comet follows the pattern of past apparitions, its tail will be far more developed after perihelion than before, and because of the angle at which we see it, the tail will probably appear longest in late March and early April. By this time the comet will have reentered the "dashed" part of its orbit, and will be travelling farther and farther south in the sky. This means that observers in the southern hemisphere, or at least in low northern latitudes, will enjoy the best views of the comet in its more spectacular post-perihelion phase, while northern observers will have the best view of it as a fainter object approaching the Sun.

At the 1910 apparition, conditions were generally more favorable. The month was April, the Earth lying at around longitude 210°. At perihelion, the comet lay to the west of the Sun, rising before dawn, and its tail was seen to good advantage.

Binocular observers can expect to view the comet round about the beginning of November 1985, when its total magnitude is predicted to have reached about 8. (The author should, however, insert the routine warning about trusting to prediction and routine, particularly in the matter of cometary magnitudes!) It may be followed until early summer 1986, when it should be fading from visibility in small instruments. At its maximum visible development, a nuclear magnitude of about 4, and a tail length of up to 20° (equivalent to a true length of some 50 million miles) may be expected, but the tail will be seen well only in dark country skies, with the comet well above the horizon. The table on page 157 gives its predicted position, and total magnitude, at five-day intervals over the expected period of binocular visibility.

In addition to this table, the set of diagrams following page 158 indicates the comet's position in the western sky (pre-perihelion) and the eastern sky (post-perihelion), at five-day intervals during the observable periods between the beginning of January and the end of March 1986. These diagrams have been prepared for observers in six different latitudes: 50°, 35°, and 20° north; on the equator (0°); and 20° and 40° south. They are correct for the moment when the Sun lies 12° below the horizon—the time of transition between nautical and astronomical twilight. This is approximately the time when the sky becomes too bright for useful observation before dawn, or begins to be dark enough for useful observation after sunset. The interval between this moment and the time of local sunrise or sunset depends upon the latitude and, to some extent, the season. Average values are as follows:

Latitude (N or S)	Interval after sunset or before sunrise
50°	75^m
40	60
35	55
20	50
0	45

These diagrams show clearly that the comet will be very poorly seen indeed from northern latitudes of 50° and greater in its post-perihelion phase. It should be noted that the tail length, as shown, may bear little resemblance to that observed; when the comet is within some 20° of the horizon, even the head may be hard enough to see, unless sky conditions are unusually favorable.

*The predicted position of Halley's Comet, November 1, 1985–June 20, 1986**

DATE (0^h U.T.)	R.A.	DEC.	MAG.
1985			
Nov 1	$05^h\ 22^m$	+21° 47′	8.9
6	05 00	22 06	8.4
11	04 30	22 12	7.9
16	03 50	21 46	7.3
21	02 59	20 19	6.9
26	02 02	17 33	6.5
Dec 1	01 07	13 43	6.3
6	00 19	09 40	6.1
11	23 41	06 05	6.0
16	23 12	03 11	5.9
21	22 50	+00 56	5.8
26	22 32	−00 50	5.6
31	22 18	−02 14	5.4
1986			
Jan 5	22 07	−03 24	5.2
10	21 57	04 24	4.8
15	21 47	05 19	4.5
20	21 39	06 13	4.1
25	21 30	−07 07	3.7
Feb 20	20 43	−13 19	3.6
25	20 34	14 51	4.3
Mar 2	20 26	16 34	4.8
7	20 17	18 31	5.0
12	20 06	20 50	5.0
17	19 53	23 41	4.8
22	19 36	27 22	4.5
27	19 09	32 15	4.3
Apr 1	18 24	38 36	4.1
6	17 03	45 16	4.0
11	14 58	47 09	4.0
16	13 03	40 30	4.4
21	11 54	31 12	4.9
26	11 16	23 41	5.5
May 1	10 55	18 22	6.1
6	10 42	14 38	6.5
11	10 34	11 57	6.9
16	10 29	10 01	7.3
21	10 26	08 35	7.6
26	10 24	07 30	7.8
31	10 24	06 42	8.1
Jun 5	10 24	06 06	8.3
10	10 25	05 40	8.6
15	10 27	05 21	8.8
20	10 28	−05 09	9.1

*The material for this table, together with some other information used in this section, was taken from *The Comet Halley Handbook* by Donald K. Yeomans, published in 1981 by NASA.

The diagrams on the following pages show the position of Halley's Comet in the evening and morning sky, for different observer latitudes. The azimuth scale, along the bottom of each view, is in degrees measured in either the east or west direction, starting at the southern point of the horizon. The comet's altitude, in degrees, is also shown.

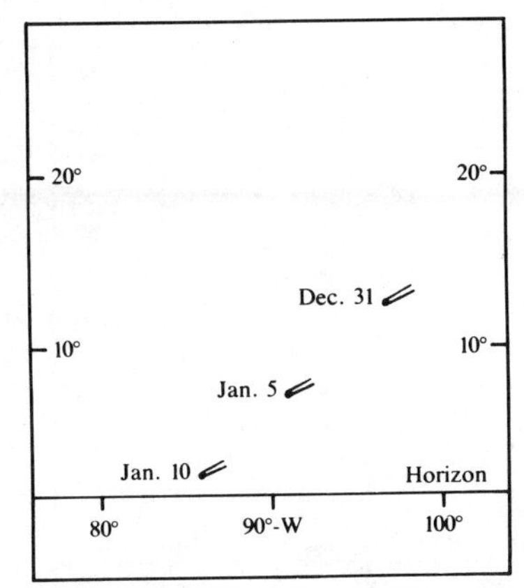

Latitude 40°S: Evening Sky

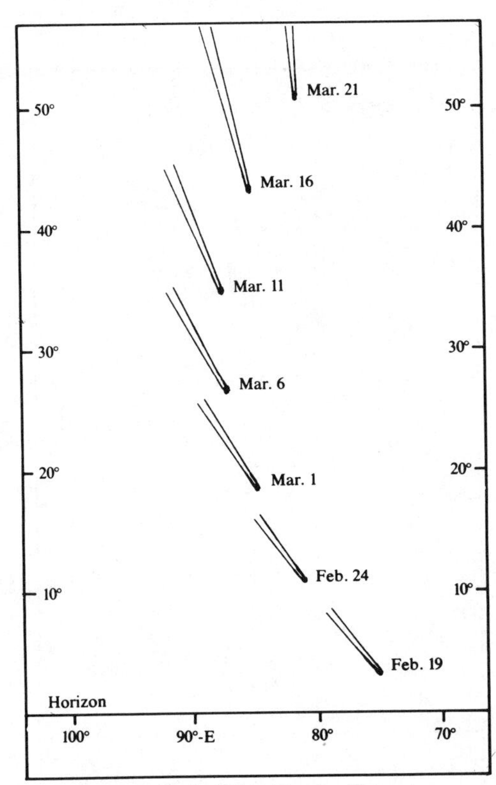

Latitude 40°S: Morning Sky

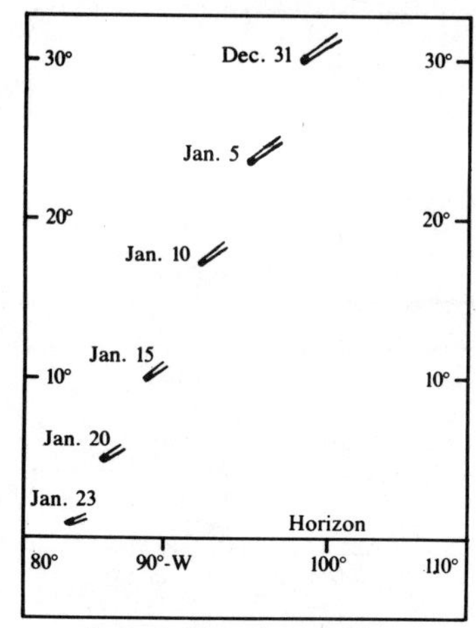

Latitude 20°S: Evening Sky

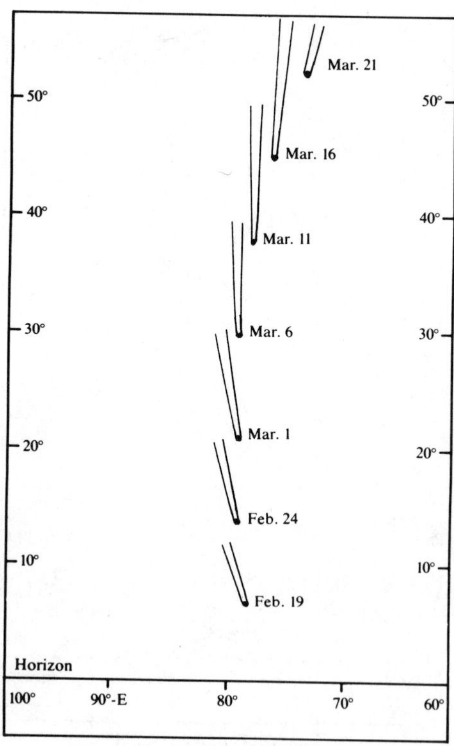

Latitude 20°S: Morning Sky

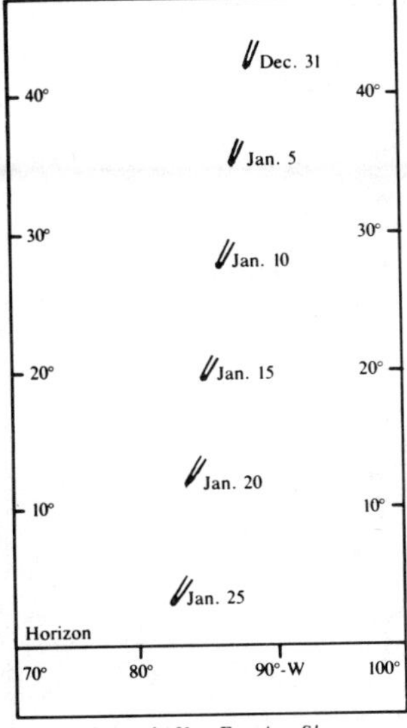

Latitude 0°: Evening Sky

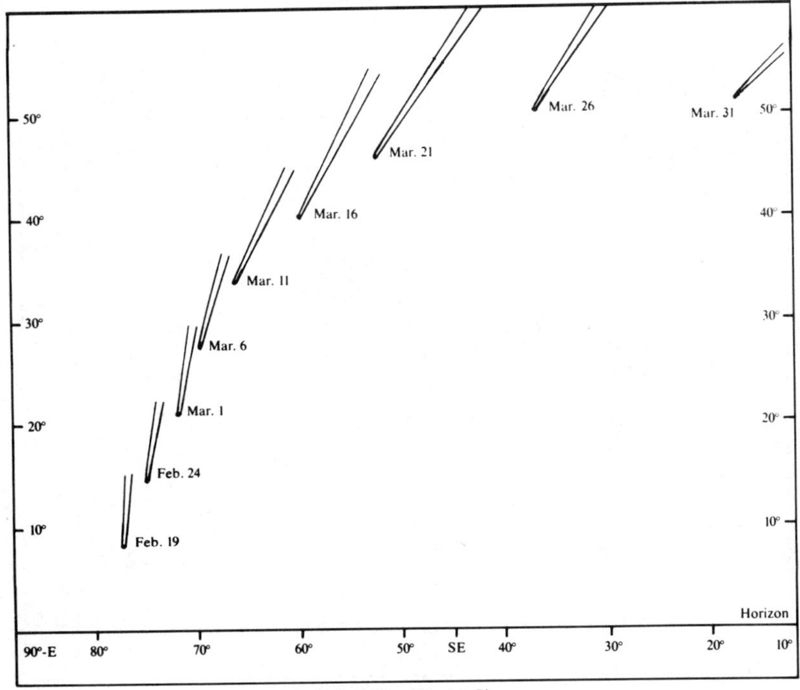

Latitude 0°: Morning Sky

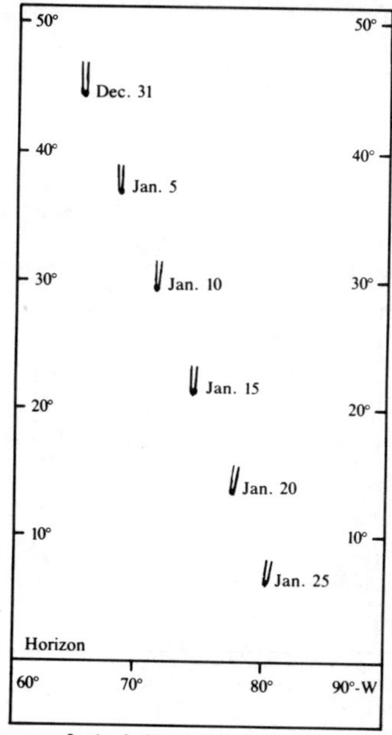

Latitude 20°N: Evening Sky

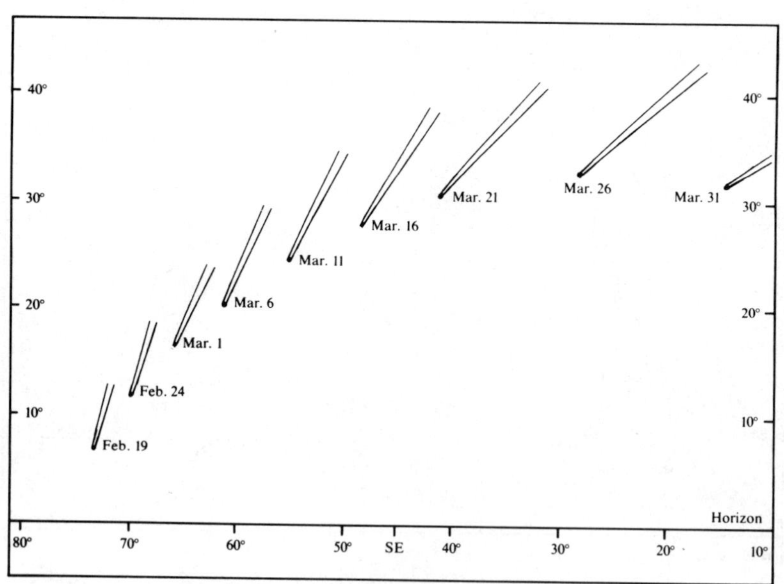

Latitude 20°N: Morning Sky

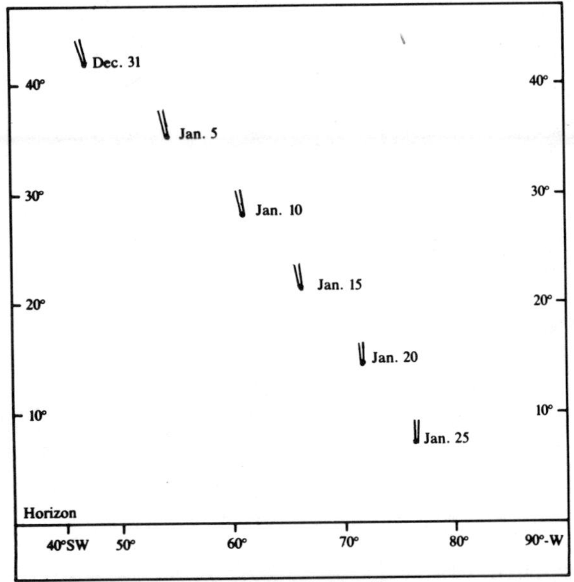

Latitude 35°N: Evening Sky

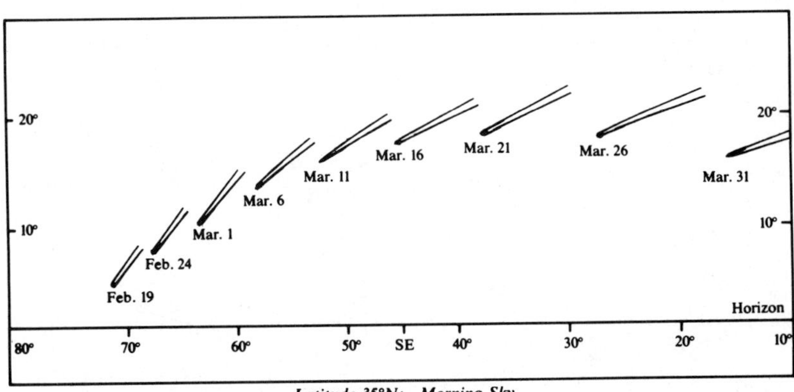

Latitude 35°N: Morning Sky

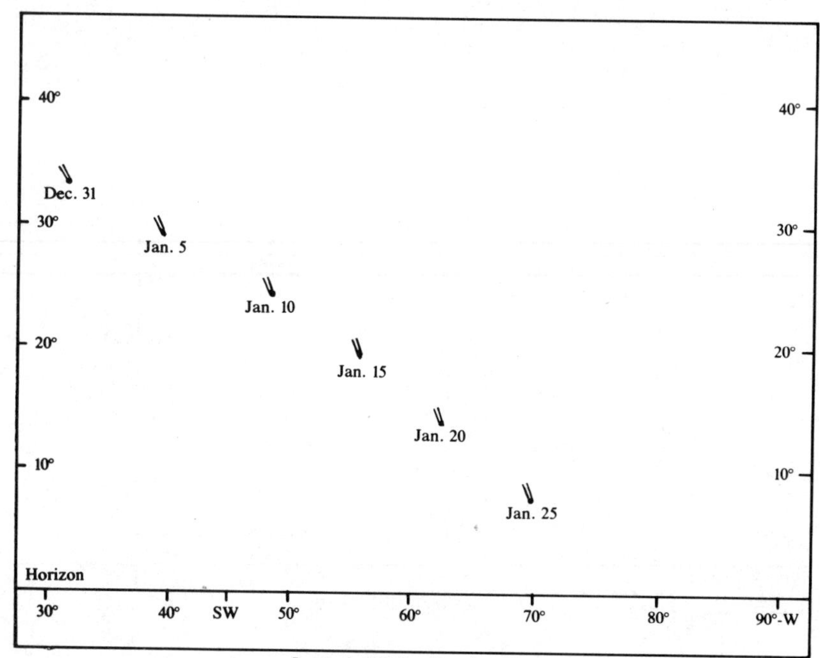

Latitude 50°N: Evening Sky

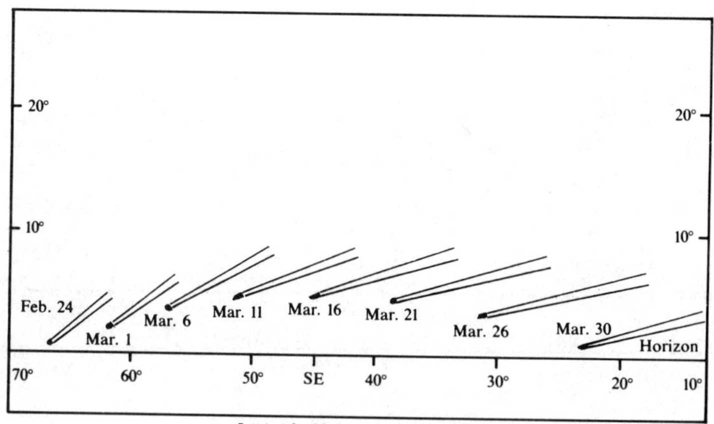

Latitude 50°N: Morning Sky

Index

(With the exception of the popular names of certain stars and other well-known objects, the individual objects listed by constellation in Chapter 7 are not included.)